U0157743

湖南大学建筑与规划学院教学成果丛书

设计的检验 理性与创新

湖南大学建筑与规划学院优秀毕业设计汇编

2015-2021

Compilation of Graduation Design of School of
Architecture and Planning, Hunan University

湖南大学建筑与规划学院教学成果编写组 编

中国建筑工业出版社

图书在版编目（CIP）数据

设计的检验　理性与创新：湖南大学建筑与规划学院优秀毕业设计汇编：2015-2021 = Compilation of Graduation Design of School of Architecture and Planning, Hunan University / 湖南大学建筑与规划学院教学成果编写组编. -- 北京：中国建筑工业出版社，2022.9

（湖南大学建筑与规划学院教学成果丛书）

ISBN 978-7-112-27762-9

Ⅰ. ①设… Ⅱ. ①湖… Ⅲ. ①建筑设计－作品集－中国－现代 Ⅳ. ①TU206

中国版本图书馆CIP数据核字(2022)第147481号

责任编辑：陈夕涛　李东　徐昌强
责任校对：王烨

湖南大学建筑与规划学院教学成果丛书

设计的检验　理性与创新
湖南大学建筑与规划学院优秀毕业设计汇编
2015-2021
Compilation of Graduation Design of School of Architecture and
Planning, Hunan University

湖南大学建筑与规划学院教学成果编写组　编

*

中国建筑工业出版社出版、发行（北京海淀三里河路9号）
各地新华书店、建筑书店经销
北京富诚彩色印刷有限公司印刷

*

开本：787毫米×1092毫米　1/16　印张：16　字数：488千字
2022年11月第一版　　2022年11月第一次印刷
定价：158.00元
ISBN 978-7-112-27762-9
（39019）

湖南大学建筑与规划学院教学成果丛书编委会

顾　问：魏春雨
主　编：徐　峰
副主编：袁朝晖　焦　胜　卢健松　叶　强　陈　翚　周　恺

设计的起点 认知与启蒙
湖南大学建筑与规划学院优秀基础教学成果汇编 2015-2021
执行主编：钟力力
参与编辑：胡梦倩　陈瑞琦　林煜芸　亓宣雯

设计的生成 过程与教学
湖南大学建筑与规划学院优秀课程设计汇编 2015-2021
执行主编：许昊皓
参与编辑：李　理　齐　靖　向　辉　邢书舟　王　蕾　刘　晴　刘　骞　杨赛尔　高美祥
　　　　　王　文　陆秋伶　谭依婷　燕良峰　尹兆升　吕潇洋

设计的检验 理性与创新
湖南大学建筑与规划学院优秀毕业设计汇编 2015-2021
执行主编：杨　涛　姜　敏
参与编辑：王小雨　黄龙颜　王　慧　李金株　张书瑜　闫志佳　叶　天　胡彭年

设计的实践 转译与传承
湖南大学建筑与规划学院优秀实践案例汇编 2015-2021
执行主编：沈　瑶
参与编辑：张　光　陈　娜　黎璟玉　廖静雯　林煜芸　陈瑞琦　陈偌晰　刘　颖　欧阳璐

设计的理论 在地与远方
湖南大学建筑与规划学院优秀研究论文汇编 2015-2021
执行主编：沈　瑶
参与编辑：何　成　冉　静　成逸凡　张源林　廖土杰　王禹廷

总体介绍

学校概况

湖南大学办学历史悠久、教育传统优良，是教育部直属全国重点大学，国家"211工程""985工程"重点建设高校，国家"世界一流大学"建设高校。湖南大学办学起源于公元976年创建的岳麓书院，始终保持着文化教育教学的连续性。1903年改制为湖南高等学堂，1926年定名为湖南大学。目前，学校建有5个国家级人才培养基地、4个国家级实验教学示范中心、1个国家级虚拟仿真实验教学中心、拥有8个国家级教学团队、6个人才培养模式创新实验区；拥有国家重点实验室2个、国家工程技术研究中心2个、国家级国际合作基地3个、国家工程实验室1个；入选全国首批深化创新创业教育改革示范高校、全国创新创业典型经验高校、全国高校实践育人创新创业基地。

学院概况

湖南大学建筑与规划学院的办学历史可追溯到1929年，著名建筑学家刘敦桢、柳士英在湖南大学土木系中创办建筑组。90余年以来，学院一直是我国建筑学专业高端人才培养基地。学院下设两个系、三个研究中心和两个省级科研平台，即建筑系、城乡规划系，地方建筑研究中心、建筑节能绿色建筑研究中心、建筑遗产保护研究中心、丘陵地区城乡人居环境科学湖南省重点实验室、湖南省地方建筑科学与技术国际科技创新合作基地。

办学历程

1929年，著名建筑学家刘敦桢在湖南大学土木系中创办建筑组。

1934年，中国第一个建筑学专业——苏州工业专门学校建筑科的创始人柳士英来湖南大学主持建筑学专业。柳士英在兼任土木系主任的同时坚持建筑学专业教育。

1953年，全国院系调整，湖南大学合并了中南地区各院校的土木、建筑方面的学科专业，改名"中南土木建筑学院"，下设营建系。柳士英担任中南土建学院院长。

1962年，柳士英先生开始招收建筑学专业研究生，湖南大学成为国务院授权的国内第一批建筑学研究生招生院校之一。

1978年，在土木系中恢复"文革"中停办的建筑学专业，1984年独立为建筑系。

1986年，开始招收城市规划方向硕士研究生。

1995年，在湖南省内第一个设立五年制城市规划本科专业。

1996年至2004年间，三次通过建设部组织的建筑学专业本科及研究生教育评估。

2005年，学校改建筑系为建筑学院，下设建筑、城市规划、环境艺术3个系，建筑历史与理论、建筑技术2个研究中心和1个实验中心。2005年，申报建筑设计及其理论博士点，获得批准。同年获得建筑学一级学科硕士点授予权。

2006年设立景观设计系，2006年成立湖南大学城市建筑研究所，2007年成立湖南大学村落文化研究所。

2008年，城市规划本科专业在湖南省内率先通过全国高等学校城市规划专业教育评估。

2010年12月，获得建筑学一级学科博士点授予权，下设建筑设计及理论、城市规划与理论、建筑历史及理论、建筑技术及理论、生态城市与绿色建筑五个二级学科方向。

2010年，将"城市规划系"改为"城乡规划系"。

2011年，建筑学一级学科对应调整，申报并获得城乡规划学一级学科博士点授予权。

2012年，城乡规划学本科（五年制）、硕士研究生教育通过专业教育评估。

2012年，获得城市规划专业硕士授权点。

2012年，教育部公布的全国一级学科排名中，湖南大学城乡规划学一级学科为第15位。

2014年，设立建筑学博士后流动站。

2016年，城乡规划学硕士研究生教育专业评估复评通过，有效期6年。

2017年，在第四轮学科评估中为B类（并列11位）。

2019年，建筑学专业获批国家级一流本科专业建设点，建成湖南省地方建筑科学与技术国际科技创新合作基地；

2020年，城乡规划专业获评国家级一流本科专业建设点。

2020年，建成丘陵地区城乡人居环境科学湖南省重点实验室。

2021年，"建筑学院"更名为"建筑与规划学院"。

建筑学专业介绍

一、学科基本情况

本学科办学 90 余年以来，一直是我国建筑学专业的高端人才培养基地。1929 年，著名建筑学家刘敦桢、柳士英在湖南大学土木系中创办建筑组；1953 年改为"中南土木建筑学院"，成为江南最强的土建类学科；1962 年成为国务院授权第一批建筑学专业硕士研究生招生单位；1996 年首次通过专业评估以来，本科及硕士研究生培养多次获"优秀"通过；2011 年获批建筑学一级学科博士授予权；2014 年获批建筑学博士后流动站；2019 年获批国家级一流本科专业建设点。

二、学科方向与优势特色

下设建筑设计及理论、建筑历史与理论、建筑技术科学、城市设计理论与方法 4 个主要方向，通过科研项目和社会实践，实现前沿领域对接，已形成了"地方建筑创作""可持续建筑技术""绿色宜居村镇""建筑遗产数字保护技术"等特色与优势方向。

三、人才培养目标

承岳麓书院千年文脉，续中南土木建筑学院学科基础，依湖南大学综合性学科背景，适应全球化趋势及技术变革特点，着力培养创新意识、文化内涵、工程实践能力兼融的建筑学行业领军人才。

城乡规划专业介绍

一、学科基本情况

本学科是全国较早开展规划教育的大学之一，具有完备的人才培养体系（本科、学术型/专业型硕士研究生、学术型/工程类博士、博士后），湖南省"双一流"建设重点学科。本科和研究生教育均已通过专业评估，有效期 6 年。

二、学科方向与优势特色

学位点下设城乡规划与设计、住房与社区建设规划、城乡生态环境与基础设施规划、城乡发展历史与遗产保护规划、区域发展与空间规划 5 个主要方向，通过科研项目和社会实践，实现前沿领域对接，已形成了城市空间结构、城市公共安全与健康、丘陵城市规划与设计、乡村规划、城市更新与社区营造等特色与优势方向。学科建有湖南省重点实验室"丘陵地区城乡人居环境科学"、与湖南省自然资源厅共建"湖南省国土空间规划研究中心"、与住房与城乡建设部合办"中国城乡建设与社区治理研究院"。

三、人才培养目标

学科聚焦世界前沿理论，面向国家重大需求，面向人民生命健康，服务国家和地方经济战略，承担国家级科研任务，产出高水平学术成果，提供高品质规划设计和咨询服务，在地方精准扶贫与乡村振兴工作中发挥作用，引领地方建设标准编制，推动专业学术组织发展。致力于培养基础扎实、视野开阔、德才兼备，具有良好人文素养、创新思维和探索精神的复合型高素质人才。

Introduction to Hunan University

Hunan University is an old and prestigious school with an excellent educational tradition. It is considered a National Key University by the Ministry of education, is integral to the national "211 Project" and "985 Project", and has been named a national "world-class university". Hunan University as it is today, originally known as Yuelu Academy, was founded in 976 and has continued to maintained the culture, education, and teaching for which it was so well known in the past. It was restructured into the university of higher education that exists today in 1903 and officially renamed Hunan University in 1926. The university has five national talent training bases, four national experimental teaching demonstration centers, one national virtual simulation experimental teaching center, eight national teaching teams, and six talent training mode innovation experimental areas. The school is also well equipped in terms of facilities, as it has two national key laboratories, two national engineering technology research centers, three national international cooperation bases, and one national engineering laboratory. It has also received many honors, as it is considered one of the top national demonstration universities for deepening innovation and entrepreneurship education reform, one of the top national universities with opportunities in innovation and entrepreneurship, and one of the top national universities' for practical education, innovation, and entrepreneurship.

School overview

The origin of the School of Architecture and Planning at Hunan University can be traced back to 1929, when famous architects Liu Dunzhen and Liu Shiying founded the architecture group as part of the Department of Civil Engineering. For more than 90 years, it has been a high-level talent training base for architecture in China. The school has two departments, three research centers, and two provincial scientific research platforms, namely, the Department of Architecture, the Department of Urban and Rural Planning, the Local Building Research Center, the Energy-saving Green Building Research Center, the Building Heritage Protection Research Center, the Hunan Provincial Key Laboratory of Urban and Rural Human Settlements and Environmental Science in Hilly Aeas, and the Hunan Provincial Local Science and Technology, International Scientific and Technological Innovation Cooperation Base.

Timeline of the University of Hunan's development

In 1929, the famous architect Liu Dunzhen founded the construction group within the Department of Civil Engineering at Hunan University. In 1934, Liu Shiying, the founder of the Architecture Department of the Suzhou Institute of Technology, which was the first one to provide major in architecture in China, came to Hunan University to preside over architecture major. Liu Shiying insisted on architectural education while concurrently serving as the director of the Department of Civil Engineering.

In 1953, with the adjustment of national colleges and departments, Hunan University merged their disciplines of civil engineering and architecture with various colleges and universities in central and southern China, forming a new institution that was renamed "Central and Southern Institute of Civil Engineering and Architecture". At this new institution, they set up a Department of Construction. Liu Shiying served as president of the Central South Civil Engineering College.

In 1962, Liu Shiying began to recruit postgraduates majoring in architecture. Hunan University became one of the first institutions authorized by the State Council to recruit postgraduates in architecture in China.

In 1978, Liu Shiying resumed providing the architecture major in the Department of Civil Engineering, which had been suspended during the Cultural Revolution. The Department of Architecture became independent in 1984.

In 1986, the University of Hunan began to recruit master's students to study urban planning.

In 1995, the first five-year official undergraduate major in urban planning was established in Hunan Province.

From 1996 to 2004, the university passed the undergraduate and graduate education evaluation of architecture organized by the Ministry of Construction three times.

In 2005, the school changed its architecture department into an Architecture College, which included the three departments of architecture, urban planning, and environmental art design, two research centers for architectural history, theory, and architectural technology respectively, and one experimental center.In 2005, the university applied to provide a doctoral program of architectural design and theory, which was approved. In the same year, it was also granted the right to provide a master's degree in architecture.

In 2006, the Department of Landscape Design and the Institute of Urban Architecture at Hunan University were established.In 2007, the Institute of Village Culture at Hunan University was established.

In 2008, the undergraduate major of urban planning took the lead in passing the education evaluation for urban planning majors in national colleges and universities in Hunan Province.

In December 2010, Hunan University was granted the right to provide a doctoral program in the first-class discipline of architecture, with five second-class discipline directions, including: Architectural Design and Theory, Urban Planning and Theory, Architectural History and Theory, Architectural Technology and Theory, and Ecological City and Green Building Design.

In 2010, the "Urban Planning Department" was changed to the "Urban and Rural Planning Department".

In 2011, the university applied for and obtained the ability to transform the first-class discipline of architecture to provide the right to grant the doctoral program of the first-class discipline of urban and rural planning.

In 2012, the undergraduate (five-year) and master's degree in education in urban and rural planning passed the professional education evaluation.

In 2012, it obtained the authorization to provide a master's in urban planning.

In the national first-class discipline ranking released by the Ministry of Education in 2012, the first-class discipline of urban and rural planning of Hunan University ranked 15th overall.

In 2014, a post-doctoral mobile station for architecture was established.

In 2016, the degree program for a Master of Urban and Rural Planning was given a professional re-evaluation and passed, which is valid for another 6 years.

In 2017, the university was classified as Class B and tied for 11th place in the fourth round of discipline evaluation.

In 2019, the architecture specialty was approved as a National First-Class Undergraduate Specialty Construction Site and built into an international scientific and technological innovation and cooperation base of local building science and technology in Hunan Province.

In 2020, the major of urban and rural planning was rated as a national first-class undergraduate major construction point.

In 2020, the school began construction on the Hunan Key Laboratory of Urban and Rural Human Settlements and Environmental Science in Hilly Areas.

In 2021, the "School of Architecture" was renamed the "School of Architecture and Planning".

Introduction to architecture

1. Discipline overview

This university has provided a high-level talent training base for architecture in China for more than 90 years. In 1929, famous architects Liu Dunzhen and Liu Shiying founded the construction group in the Department of Civil Engineering of Hunan University. In 1953, the department was transformed into the Central South Institute of Civil Engineering and Architecture, becoming the leading institute in the Southern Yangzi River (Jiangnan). In 1962, the program was among the first graduate enrollment units of architecture authorized by the State Council. Since passing the professional evaluation for the first time in 1996, the cultivation of undergraduate and postgraduate students has maintained the grade of "excellent" in the many following evaluations. In 2011, the university was granted the right to provide a doctorate degree of the first-class discipline of architecture. The department was approved as a post-doctoral mobile station in architecture in 2014. In 2019, it was approved as a national first-class undergraduate professional construction site.

2. Discipline orientation and features

The degree of Architecture at Hunan Uni versity has four main academic directions: Architectural Design and Theory, Architectural History and Theory, Architectural Technology Science, and Urban Design Theory and Methods. Through scientific research projects and social practice, school has established a serial of featured fields, which include "Local Architectural Creation and Praxis", "Sustainable Architectural Technology", "Green Livable Villages and Towns", and "Digital Protection Technology Of Architectural Heritage".

3. Objectives of professional training

The program of degree strives to inherit the thousand-year history of Yuelu Academy, continue the discipline foundations of the Central South Institute of Civil Engineering and Architecture, follow the comprehensive discipline background of Hunan University, adapt to the trend of globalization and the characteristics of technological change, and strive to cultivate high-level leading talents of architecture for the industry with innovative thinking, high humanistic intuition, solid and broad engineering practice ability.

Introduction to urban and rural planning

1. Discipline overview

The degree program at Hunan University is among the earliest ones in China to provide planning education. It has a complete professional training system, from undergraduate, academic, and professional postgraduate programs to academic and engineering doctoral and postdoctoral programs, and it is considered a double first-class key department in Hunan Province. Both the undergraduate and graduate education tracks have passed professional evaluation and are valid for 6 years.

2. Discipline orientation and features

This degree program includes five academic areas: Urban and Rural Planning and Design, Housing and Community Construction Planning, Urban and Rural Ecological Environment and Infrastructure Planning, Urban and Rural Development History and Heritage Protection Planning, and Regional Development and Spatial Planning. Through scientific research projects and social practice, the program has established a serial of featured fields, and provides curriculums for urban spatial structure, urban public safety and health, hilly urban planning and design, rural planning, urban renewal, and community construction. The program provides access to the Hunan Key Laboratory on the Science of Urban and Rural Human Settlements in Hilly Areas, the Hunan Provincial Land and Space Planning Research Center that was jointly built with Hunan Provincial Department of Natural Resources, and the China Academy of Urban and Rural Construction and Social Governance which was jointly organized with the Ministry of Housing and Urban Rural Development.

3. Objectives of professional training

The program focuses on the cutting-edge theories, tackles major national needs and the problems surrounding individual quality of life, serves national and local economic strategies, undertakes national scientific research tasks, produces high-level academic achievements, provides high-quality planning, design, and consulting services, plays a role in local targeted poverty alleviation and rural revitalization, leads the preparation of local construction standards, and promotes the development of professional academic organizations. We are committed to cultivating high-caliber talents with a solid educational foundation, broad vision, political integrity and talent, high moral compass, innovative thinking abilities, and exploratory spirit.

建筑学专业毕业设计总体介绍

建筑学专业毕业设计建设情况

1. 毕业设计选题

湖南大学建筑学专业教学组织过程注重毕业设计。2011年以来，逐年对毕业设计的要求做进一步强化，每年召开专题研讨会，研究命题的科学性。已逐步形成历史建筑虚拟修复设计、绿色高层建筑设计、校园建筑设计、公共文化建筑设计、新四校联合毕设、西南四校联合毕设、"旧城栖居"联合毕设等较稳定的特色选题方向。

毕业设计选题遵循以下原则：规模在一万平方米左右（高层建筑可放宽要求），不宜过小以免降低难度，使其复杂性不足；也不宜过大以免增加学生工作量，影响设计成果深度。毕业设计命题均以实际用地，结合实际工程要求适当调整经济技术指标，课题接近实际工程条件；毕业设计成果要求建筑设计部分达到初步设计深度要求，另需提交设计前期研究文案、设计论文、主要楼层结构构造详图等图纸，选题的内容、难度和综合性均应高于课程设计，能涵盖本科教育中的主要知识点。

2. 毕业设计指导

2008~2010年，毕业设计由副教授以上职称，或工程经验丰富的教师指导完成，每位教师指导8名学生；2011年之后，按照学校要求，每位教师均需参加毕业设计指导，我院因此组织了不同毕业设计小组指导毕业设计，平均每位教师指导3名学生；2013年后，组织了毕业设计命题筛选活动，减少命题数量，提高命题质量，以同一命题为小组指导学生毕业设计。此外，专门组织技术教学组，对毕业设计中的结构、暖通、水电专业知识进行指导，确保毕业设计组织有足够的讲师或讲师以上的教师以及相关专业的教师进行指导。教学组织中，教师与学生每周见面不得少于两次，在教师指导下，由学生独立完成自己的设计任务，提交完备的毕业设计文件。在毕业设计教学组织中，强化过程控制与中期检查的重要性，避免学生对设计进度把握不足影响设计成果。

3. 毕业设计量化评分及反馈机制

毕业设计既是本科学生培养的重要环节，也是检验建筑学教学质量的重要阶段。我院自2012年开始，根据教学计划中的课程组成，确定各个知识模块的权重，实施毕业设计量化评分尝试；进而应用各个知识板块的得分情况，对本科教学计划实施中的薄弱环节予以判定，形成设计教学的反馈机制，推动教学改革有目的的实施。

严格控制毕业设计深度要求，严格把控毕业设计的评分过程。设计评分实施了量化评分制度，避免各个答辩小组评分不均的情况；各个答辩小组包含建筑设计、城乡规划、建筑技术专业教师，以及外聘设计院专家一名；设计成绩教师占30%，答辩小组成绩占70%，构成合理；严格控制毕业设计成果，中期检查不合格者推迟答辩（暑期之后），答辩成绩不合格重修（来年随低年级重新修课），每年淘汰一定数量的不合格作业。

4. "鹿鸣赫曦"全国建筑学类建筑设计分享活动

2020年起，由湖南大学建筑与规划学院发起，全国建筑学专业近二十所兄弟院校共同举办了题为"鹿鸣赫曦"的年度建筑学类毕业设计分享活动（含建筑设计、城市设计、建筑遗产保护与更新等方向），为青年学子、建筑学界提供了一个毕业设计的交流平台。活动采用线上方式，分享下列内容：1）毕业设计组织、命题心得，通过交流提高毕业设计命题质量；2）优秀的学生代表、毕业设计创作作品，以榜样激发学生的兴趣与创造力。此外，活动还邀请来自各大设计机构及院校的专家作为评论嘉宾，与参与分享的师生展开对谈与研讨。活动至今已成功举办两届，在学界引起了较热烈的反响，活动参与人数近2万人次，为全国建筑学类毕业设计注入了新的活力。

鹿鸣赫曦·2020 年度建筑学类毕业设计分享会

鹿鸣赫曦·2021 年度建筑学类毕业设计分享会

课题一

历史建筑虚拟修复设计

出题人：张卫、肖灿、钟明芳、欧阳虹彬

一、设计主题

通过历史建筑虚拟修复研究性设计，理解建筑遗产保护的基本原则，较为熟练地掌握历史建筑保护中史料调研、分析及复原推理与求证的方法，并运用相关软件进行虚拟展示，基本掌握建筑虚拟修复的方法与技能。对所选历史建筑的文化、技术及艺术价值有较为全面的梳理，对其建筑的整体和细部均有严谨的考证与推断，虚拟修复展示系统逼真。

二、设计内容

每位同学选择一栋（或一组）大部分或全部被损毁的历史建筑，对其历史资料进行收集整理与分析，并在考证其相关历史文化背景、图纸、数据和科学推断的基础上，将该历史建筑用相关软件在数字平台上完整修复并全面展示，并可结合一定的遗址更新利用概念设计提出数字复原展示方案。

三、设计要求

1. 历史文化考据

撰写 5000~8000 字关于所选历史建筑虚拟复原的历史文化考证与推断依据论文一篇。

2. 设计成果

(1) 设计总说明（包括所选历史建筑的独特性与代表性、历史沿革、考证与推断依据，虚拟修复展示方案的可行性，如有更新利用概念设计则说明其设计的理念与目标等）

(2) 所选历史建筑现状的总平面图纸（区域位置图、场地现状图）

(3) 分析图纸（建筑历史分析、建筑形式修复分析、数字建筑展示方式等）

(4) 建筑复原修复图纸（建筑复原修复平面图、立面图、剖面图、大样图、建筑整体及重点空间室内场景还原效果图）

(5) 技术图纸（被修复建筑的细部及代表性节点，不少于2个）

天心阁实拍图

天心阁实拍图

清末天心阁总图关系

民国天心阁总图关系

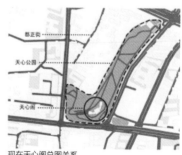

现在天心阁总图关系

课题二

历史建筑虚拟修复与更新系统设计
出题人：张卫、肖灿、钟明芳、欧阳虹彬

一、设计主题

通过该研究性设计，理解建筑遗产保护的基本原则，较为熟练地掌握历史建筑保护中史料调研与分析的方法，并运用相关软件进行虚拟展示，基本掌握建筑虚拟修复的方法与技能。对所选历史建筑的文化、技术及艺术价值有较为全面的梳理，对其建筑的整体和细部均有较严谨的考证与推断，虚拟修复展示系统逼真，更新方案需要体现对历史建筑的可持续发展利用的理念（包括空间、功能、材料、节能设计等）。

二、设计内容

每位同学选择一栋（或一组）部分或全部被损毁的历史建筑，通过对其历史资料的收集整理与分析，将该历史建筑用相关软件在数字平台上完整修复并全面展示，同时提出一个更新利用的设计方案。

三、设计要求

1. 设计总说明

包括所选历史建筑的独特性与代表性、历史沿革、考证与推断依据、虚拟修复方案的可行性、更新利用设计的理念与目标等。

2. 设计成果

(1) 所选历史建筑现状的总平面图纸（区域位置图、场地现状图）

(2) 分析图纸（建筑历史分析、建筑形式修复分析、建筑更新分析等）

(3) 建筑修复图纸（建筑修复平面图、立面图、剖面图及大样图、建筑更新利用的总平面图、平面图、立面图、剖面图）

(4) 技术图纸（被修复建筑的细部及代表性节点，不少于2个）

(5) 虚拟修复的动态展示成果（对历史建筑外观、空间、结构及其与场地等的关系进行全面、系统展示）

区位图

图书馆内部

图书馆实拍图

图书馆实拍图

课题三

城市建筑绿色设计与预评
长沙绿色高层毕业设计

出题人：徐峰、刘宏成、谢菲、何成

一、设计主题

当前，中国经济发展需要建设与示范具有绿色低碳观念的文创办公城市空间，鼓励将最新的绿色低碳建筑技术及其研究成果转化运用到建筑设计中，这些是当前城市生态绿色文明建设的重要步骤，也是实现城市发展转型的具体表现。因此，在城市化进程中顺应大众福祉的国家发展策略，必然需要创建凸显时代建筑人文特性和永续价值观的公共资源，建设服务于文化创意产业、宣传与培育文化发展成果的媒体设施，是长沙的城市发展战略。

二、设计内容

2014 年"长沙市城市总体规划"将设计项目所处的地段规划到"新世纪片区分区规划大纲"中，并将基地所在的马栏山片区功能定位为文化娱乐、文教旅游和居住为主的综合片区（原来以居住用地、防护绿地为主）。目前，此长沙绿色高层（现中标方案名为湖南创意设计总部大厦）拟用地为城市建设用地，规划用途修改论证为商务和绿化，因此急需完善配套设施，提升景观环境。基地毗邻东二环高架，周边有建成的滨河路等道路系统。

三、设计要求

项目建设基地是地块 18（X06－A49）、地块 19（X06-A56-1）。容积率建议为 3.0 左右（建议不大于

3.5），可以根据自拟任务书调整。耐火等级一级，抗震设计 6 度抗震设防考虑。建筑类型为公共建筑。建筑的功能空间组成自行拟定。

设计主要分成两个考查内容：其一，总体建筑设计方案；其二，绿色建筑设计和技术路径。

总体设计方案包括两个阶段设计：首先是设计结合场地城市环境的建筑群体规划和空间设计策略，其次是绿色建筑高层／多层设计（可以参考四年级课程设计过程），需要在专项设计分析下进行深入的绿建设计及自评估。

区位图

区位图

区位图

课题四

湖南创意文化设计总部大厦
绿色建筑设计

出题人：徐峰、刘宏成、谢菲、何成

一、设计主题

两栋商业办公相结合的综合体大厦，服务于新媒体工作者、创客、研发工作者以及外来游客。以健康桥、社区桥、文化桥和培养桥为媒介打造富有活力的办公空间，室外的公共平台与屋顶花园创造了人与人交往的可能，阶梯式绿化休息空间也提供了优美的景观和适宜的微气候，同时室内跑道和健身房也为使用者提供了更为健康的生活方式。

二、设计内容

1. 基于前期场地环境调研、总体绿建设计策略、场地建筑群空间形态，确定单体建筑及其设计策略。
2. 拟定概念设计前期分析报告。
3. 自拟绿色建筑单体设计任务书。
4. 建筑单体设计。
5. 阶段性建筑比例模型性能分析。
6. 建筑低碳绿色建造技术选择和应用设计。
7. 建筑总体方案深化和设计展示。

三、设计要求

1. 场地设计

总平面图：建筑物、广场、停车场、高程、道路、出入口位置。

2. 建筑设计

设计图纸：平面图、立面图、剖面图、2~3张详图、防火分区示意图。

结构设计：结构造型。

设计效果图：表现主体建筑和重要节点。

绿色建筑设计专题：至少进行两项环境模拟并有相关模拟彩图和优化过程。

四、设计原则

1. 两个塔楼均需满足办公、商业、公寓的功能需求。
2. 达到绿建二星标准。
3. 对于噪声、采光和通风做出相应的优化设计。
4. 做出适应新媒体的办公模式的设计方案：促进员工之间的学术交流、为员工提供健康舒适的工作环境（辅之以医疗、健身、娱乐、学习空间）。
5. 有对市民开放的公共空间（室外开放平台、屋顶花园）。
6. 方案中加入景观设计：结合阶梯绿化、垂直绿化和屋顶花园。
7. 对建筑幕墙做出节能设计（双层玻璃幕墙、活动遮阳百叶）。

效果图

课题五

"归家之家" 社区营造
2017 届旧城区社区老年人康复中心毕业设计

出题人：张蔚、蒋甦琦

一、设计主题
本次设计旨在为社区病后需要康复的老人、失能失忆老人、孤寡老人营造提供康复治疗、护理关怀的社区老年人康复治疗中心。尽量消除机构化养老的各种弊端，依据多元化的治疗和护理模式为老年人营造多义养老生活空间——一个温馨的"归家之家"。

二、设计内容
康复医疗建筑功能流线组织及空间特性了解；康复护理中心适老化休闲、交流、等候空间营造；康复建筑景观设计；绿色建筑设计；材料运用等。

三、设计要求
1. 指标数据
建筑容积率：1
绿地率：≥ 30%
地面建筑层数：≤ 4 层

2. 设计功能
(1) 综合医疗区
a. 门诊（临床科室）
b. 健康评定科
c. 医技检验科
d. 治疗部
e. 康复病房
总共 180 床，30 床为 1 个护理单元，共 6 个护理单元。
每床净使用面积为病床的面积 + 过道的面积 + 医疗设备摆放预留面积。
另需加卫生间、可考虑家庭陪护面积及交流休息面积。
康复病房
医护办公
库房
等候、休闲交流等公共空间。
f. 行政办公

(2) 生活服务部
a. 生活服部务
b. 娱乐部

(3) 社区健康服务部
a. 社区健康咨询中心
b. 心理咨询室
c. 日间护理中心
d. 其他：任务书补齐，面积自定
其他社区设施根据问卷调查及相应的服务人群确定娱乐空间内容。

(4) 附属用房
(5) 户外空间设计

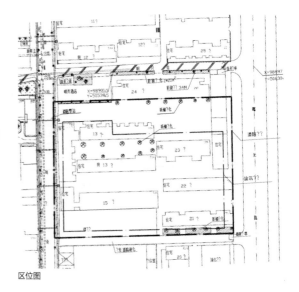

区位图

课题六

凤凰县文昌阁小学及周边有机更新设计

出题人：苗欣

一、设计主题

旅游大潮冲击下的历史城镇——凤凰，历史悠久小学校的困境与策略：名校吸引学生人数激增与校园规模的原静态界定方法的冲突，外来人口及陪读求学人员进驻片区也导致地价激增，使校区周边场地被蚕食以至阻碍校园空间的生长，原有校舍在新的教学内容调整上的不可持续……这些促成我们重新审视共享时代旅游村镇历史名校的空间再组织，从学校内外、周边以至区域上审视当下这一广泛存在的问题，基于从校园人群、陪读人群、服务人群以及旅游人群的冲突中寻求方向，给出合理的校园空间组织。

二、设计内容

1. 各建筑单体面积规划

文昌阁本部基础教学区的梳理以及改扩建：可自拟设计任务书。

文华山山脚校内自然景观区扩建：通用空间多功能活动场地（可临时搭建并便于拆除，以应对学校各种活动）。

近虹桥西路扩建区：综合服务楼（包含拓展教学、生活服务、社区学校共建三部分）。通过对山地地形处理和种植平台的设置，合理布置车道，同时使建筑肌理与古城社区肌理相协调。

2. 运动场地要求

因地制宜，充分利用地形结合当地特色项目设置相应体育活动，梳理原有运动场地，相应增加：传统武术与花鼓舞练习兼展示场地1个、跑道1条（可考虑与邻近老箭道坪小学运动场共建，不做强行要求）。

3. 城市设计层面思考

精神防线的构建：通过改扩建之后校园整体态势和学校和社区的共建关系，以悠久的学校历史和稳定的社区记忆带来的力量在有限的范围内抵御古城景区商业文明扁平化的扩张。

可持续的过渡状态的考虑：扩建部分考虑既是满足现在日益增长的需求状态，同时也考虑作为一个过渡时期的状态，为未来的改建、扩建做打算。

三、设计要求

前期调研；基地分析（区位、空间结构、社区、交通、景观、风、日照辐射、文脉等）；案例分析；建筑设计；设计完成深度、时间节点与进度详见毕业设计统一要求；各类分析图；景观环境设计。

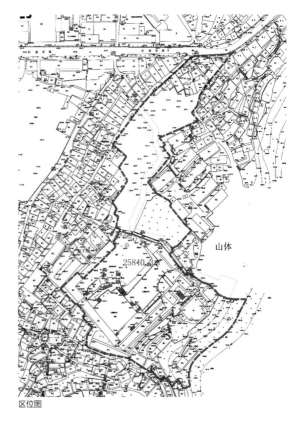

山体

25840.

区位图

课题七

长沙市工人文化宫更新计划与设计

出题人：蒋甦琦、张蔚、向昊

一、设计主题

文化宫是特殊历史时期的产物，与工人社区联系紧密，其存在与当下社会的文化需求有所区别。如何维持公益性，维持文化的自更新自生产，提供融入地域生活圈的新型文化建筑，是本次设计的探索方向。

场地内的水塘是东塘地区的名称来源，是地方文脉的重要传承，绿树环绕，宁静优美，故争取将其保留，并使自然融入设计，为建筑创造良好的自然环境。

二、设计内容

长沙市工人文化宫更新改造工程位于东塘砂子塘街区。根据前期调研、区域城市设计和概念设计指导，拟建设一个微型公寓与文化建筑的混合体。场地内同时规划有其他建筑：南侧城市综合体的建设将为场地带来稳定的人流，结合南侧的巴比伦溜冰城，以及东侧大学，设置歌舞娱乐场所、夜市、文创设施等，一定程度上恢复地域原有的特色活动，并将其转变为连接地铁站的商业后街的一部分。同时将场地内水塘部分作为洼地的特色保留，修整自然景观，并引导商业人流穿越连接至北部商贸城，联通门前的商业地下街，形成体验丰富的循环动线。

三、设计原则

1. 新建文化宫在设计功能上要满足群众艺术、娱乐、读书交流、休闲活动和文化活动中心的功能需求。同时，提供一定数量面向都市白领、学生、个体和自由职业者的微型公寓，形成功能混合、氛围融洽的新工人社区。

2. 新馆的主要功能空间应该是开敞且注重公益性，便于灵活互换使用，以建造多元化的交流、娱乐、学习的空间。

3. 总体布局合理，以保证服务大众文化交流的模式，既方便群众活动娱乐，又便于管理，实现消费功能、文化功能、社区功能三大功能合一、全面开放式管理模式。

4. 采取无障碍设计，执行有关国家标准和规范规定。

5. 注重智能设计，节约资源、健康环保，以自然生态作为设计的重要依据，注重保护基地的现有自然资源，提供一个投资合理、使用效率高、运行费用低的建筑设计，努力把文化宫建成一个绿色建筑。

前期分析

效果图

课题八

"汉口记忆"文化体验区设计
2021届建筑学专业四校联合毕业设计

出题方：华中科技大学

一、设计主题

本设计以"汉口记忆"为主题，以汉口民众乐园为依托，结合积庆里、清芬路、土垱街（今统一街）打造"汉口记忆"文化体验区。设计深入挖掘汉口老城区的地域文化特点，积极反思汉口老城更新改造中的不足，从而探索旧城区发展过程中如何保留"汉口记忆"，实现城市历史文脉的延续以及当代城市品质的提升。

二、设计内容

为满足社区文化要求，体现汉口丰厚的文化内涵，本设计以"汉口记忆"为主题，立足于拥有深厚历史文脉的汉口老街区，以汉口民众乐园为依托，结合积庆里、清芬路、土垱街（今统一街）打造"汉口记忆"文化体验区，试图将昔日汉口记忆中不可替代的本土文化与当代商业、文化、娱乐功能相融合，实现沉浸式剧场体验。人们在其中能够具有极强的体验性、参与性、娱乐性，沉浸式地感受"汉口记忆"中的建筑文化、街巷文化、戏曲文化、民间艺术文化以及民国风情。

设计应充分尊重汉口的历史，并因此留下浓厚的文化印记；体现现代建筑与传统街巷肌理、文脉文俗的传承与蜕变关系，充分结合周边建筑与街区现有状况，以现代文化功能的植入来激活汉口老街区的活力。选址和设计需重点考虑对历史建筑（如积庆里、民众乐园、国民政府旧址、武汉工艺大楼、东来顺）的保护与协调。建议由文化体验区、住宿休闲区、后勤与行政区等部分组成，具体内容及规模由各组自行根据调研成果拟定。具体功能应该依托民众乐园，打造沉浸式体验区，可以参考但不局限于以下内容：汉口里份文化博物馆、汉剧剧场、特色民宿、老字号小吃街等。

三、设计要求

1. 基地调研与选址论证报告

报告不少于5000字，A3文本，图文并茂。

2. 各组拟定的建筑设计任务书

任务书应包括民众乐园的改扩建部分，每组需涉及历史建筑的保护与更新设计。

3. 建筑设计成果

(1) 设计构思说明（包括主要经济技术指标、必要的基地分析及设计构思、功能等分析图）

(2) 总平面图1：500

(3) 各层平面图1：100~1：200（其中首层平面应表现环境设计，且各层平面主要空间应布置家具和绿化）

(4) 立面图1：100~1：200（不少于3个）

(5) 剖面图1：100~1：200（不少于2个）

(6) 局部构造详图1：20（不少于2个）

(7) 效果图：室内外透视各不少于1个，局部透视若干，主要透视不小于A3幅面

(8) 主要建构分析图

(9) 实体模型，比例自定

区位图

鸟瞰图

课题九

文化演绎，活力再生
2020届建筑学专业四校联合毕业设计

出题方：深圳大学

一、竞赛主题

无论怎样更新，也无论城市轮廓如何急剧地往高空增长，城市的发展永远不可能是在时空上唯我独尊的孤立片段。历史上引人入胜的都市样貌、充满活力的市井生活，无不体现其与自然的相互因借共生、对文化的吸纳融合，体现历时性的因地制宜、因时制宜的传承、修正、演变，体现发展片段在空间、时间及文化上的连续性。

因此，更新改造也好，高强度开发也罢，都必然顾及在特定时空中的生成逻辑。都市的快速发展与集聚，在面对典型传统及既有文脉时，如何共生、演变与传承，便是本次课题的目标。

二、设计内容

课题以"文化演绎，活力再生"为主题，选择深圳市坪山区坑梓新兴街城市更新单元范围的城中村，结合周边拟定的城市更新单元为背景，期望在深入调研、切身体验、专业领悟的基础上，为龙田世居的周边区域拟定、设计一座切合主题的都市活动（生活）场所，以期成为连接传统与当代的生活载体。

找出契合主题的关注点，在给定的基地区域范围，选择部分用地，寻找传统与当代、都市与自然、集聚与松散、高效与休闲的关联与共生，营造充满活力的生活场所，希望同时关注城市公共空间系统及其与建筑的互动。

提示的设施可以是：客家文化体验馆、旅游集散中心、特色客栈及民宿、文化商业街区、文教研习所、社区养老院（日间照料中心）等。

三、设计成果

1. 开题报告

包括基地调研、片区城市设计研究、用地选择论证1份（小组共同完成），图文并茂，3000~5000字，A3文本。

2. 建筑设计

(1) 设计构思说明

(2) 相关分析图（基地分析、方案比选、交通分析、空间分析、功能分析、景观分析等）

(3) 总平面图1：500

(4) 各层平面图1：100~1：200（首层平面表达周边环境，各层平面表达主要室内布置）

(5) 立面图1：100~1：200（不少于3个）

(6) 剖面图1：100~1：200（不少于2个）

(7) 局部构造详图1：20

(8) 经济技术指标

(9) 效果表现图（室内外透视图各不少于一个，局部场景透视若干）

(10) 成果图幅：展览图幅（A1，不少于3张），图册（A3文本）

(11) 成果模型：1：（100~200）

区位图

鸟瞰图

课题十

安化县黑茶文化展示与体验中心设计

出题方：湖南大学

2019 届建筑学专业四校联合毕业设计

一、设计主题

小桥小店沽酒，初火新烟煮茶。

地处环洞庭湖产茶区的安化县江南镇并不是行政意义上的历史文化名镇，但难掩其奇妙的自然风貌、璀璨的人文景观以及厚重的历史文化。这片土地氤氲在资水的灵秀之中，自唐代始，开世界黑茶先河，千年茶香不断。其茶汤透明洁净，叶底形质清新；香气浓郁清正，长久悠远沁心，一船香遍洞庭湖。千百年来所积累的制茶工艺和制茶文化，与勇武耐劳的古镇人融于一体，酝酿出独特而富有魅力的当地文化。曾经，古镇中人们主要以种茶、制茶为主，小作坊遍街可见，生活自由恬静且有艺术创造性。

二、设计内容

本次设计拟选取江南镇历史建筑良佐茶栈、德和茶行以及五福宫码头周边的一片基地，要求学生在深度调研之后，在框定的范围内选取一块建设用地，为热爱黑茶文化的人们设计一处学习、体验、修养心性的场所。

本次设计的主要目的是探寻历史古镇的文脉，通过建立新的空间环境秩序，为历史街区注入活力，提升生活品质。新的介入应遵循原有村落的尺度和空间结构，以善意的态度和适宜的手法，来激发老城镇的活力。

考察古镇聚落自发性生长的历史层积，及其顺应地形地貌和主动适应气候的合理性与可变性，以适度的"设计介入"或"设计干预"，来引导和掌控古镇聚落空间的演化进程。用材上充分考虑本土条件和历史风貌，建议就地取材或以再生材料为主，辅以必要的其他材料。

三、设计成果

1. 基地调研与选址论证报告

报告不少于 5000 字，A3 文本，图文并茂。

2. 建筑设计成果

(1) 设计构思说明（包括主要经济技术指标、必要的基地分析及设计构思、功能等分析图）

(2) 总平面图 1：500

(3) 各层平面图 1：100~1：200（其中首层平面应表现环境设计内容，且各层平面主要空间应布置家具和绿化）

(4) 立面图 1：100~1：200（不少于 3 个，与平面图比例统一）

(5) 剖面图 1：100~1：200（不少于 2 个，与平面图比例统一）

(6) 局部构造详图 1：20（不少于 2 个）

(7) 效果图：室内外透视各不少于 1 个，局部透视若干，主要透视不小于 A3 幅面

(8) 主要建构分析图

(9) 展览大图：A1 图幅（不少于 2 张，A3 方案文本一套），表现形式不限

区位图

建筑现状

建筑现状

建筑现状

建筑现状

课题十一

汉口历史街区保护更新及产业类历史建筑改造再利用设计

出题方：华中科技大学

一、设计主题

历史文化名城武汉，既有深厚的城市历史文化积淀，又有当下依托地区经济增长城市快速发展的活力。2017年11月1日，武汉以"老城新生"主题入选联合国教科文组织创意城市"设计之都"。其主旨在于：在城市高速发展过程中，武汉在经济增长效率、生态环境和文化遗产保护等方面已面临多重严峻挑战。如何通过创意设计向城市文化、经济、社会、生活、环境的深入渗透，促使历史性城市在持续发展中焕发新的活力，成为亟待解决的问题。基于此，本次四校联合毕业设计重点聚焦武汉汉口原德租界所在历史性城区，侧重城市历史街区保护更新和历史建筑改建再利用。

二、设计内容

本次联合毕业设计，须基于课题所在城市历史街区及其相邻街区的现场体验调研、史料收集整理，对历史街区空间结构、街区肌理、公共空间公共活动、环境景观等要素进行整体分析研究，提出课题所涉历史街区保护更新的空间策略；同时结合对原德租界工业遗产的价值研判、空间匹配关系论证，自行策划拟定设计任务书，对工业遗产单体建筑的空间转型进行目标定位，进而对其具体改建再利用展开介入性设计，以期达到工业遗产适应性再利用的目的。

三、设计成果要求

主要包括：方案图纸、手工模型（含过程模型及成果模型）、汇报PPT。每人完成不少于8张A1排版图纸，并完成单体设计模型（1：200~1：500）。

1. 历史街区研究分析内容应包含

(1) 历史街区演变因子分析

(2) 历史街区公共空间类型及层级关系分析

(3) 历史街区空间结构、空间肌理、空间类型分析

(4) 介入性设计前期策划论证分析

2. 工业遗产单体建筑内容应包含

(1) 工业遗产历史建筑特色空间及价值研判分析

(2) 新旧空间关联关系建立逻辑分析

(3) 工业遗产构成要素类型分析

(4) 区位、交通状况及周边建筑分析

(5) 原有空间特性及结构体系分析

3. 场地设计内容应包含

(1) 历史街区保护更新分析

(2) 方案生成过程分析

(3) 人流组织分析

(4) 竖向设计分析

(5) 景观设计分析

(6) 主要技术经济指标与设计说明

4. 单体建筑改建设计

应包含：各层平面图 1：200；2~3个剖面图 1：200；2~3个立面图 1：200；室内重点空间设计；室外重点节点景观意向；建筑整体鸟瞰图及主要方向透视图；主要技术措施及技术设计（包括2~3个平立剖节点）

5.15分钟PPT汇报文件

区位图

课题十二

湖南某高校校园规划及文化艺术中心设计　　出题人：袁朝晖、彭智谋

一、设计主题

本次设计以"未来时代的大学空间"为题，该主题立足于当代中国大学校园生活的现实并指向未来。对于"我的大学"之未来畅想，实质上是回应当今校园中的问题，集结技术和思想的力量，去创造一个更加多元、智慧、可持续的大学校园环境。本次设计旨在考察同学们以校园的现实参与者和未来建设者的身份，如何批判性思考大学和教育的本质，以及技术与人的关系；又如何以技术手段去解决现实社会与生态问题，并面向未来开放。

二、设计内容

1. 校园规划（一期）

根据地形特点及学校功能做出规划设计（小组设计），各建筑单体及辅助用房、设备用房、校门、构筑物设计建筑面积可根据后述功能需求及控制原则进行适当调整。在地面总建筑面积不增加的情况下各建筑面积允许做不大于 10% 的调整。

成果要求包括：规划总图；功能结构分析；交通组织分析；景观分析；海绵城市分析；风模拟分析；BIM 分析；日照分析。

2. 文化艺术中心单体设计

本建筑为湖南某大专院校文化艺术中心，定位为校园标志性建筑，相关配套用房可根据楼宇功能及相关规范、规定设置。

三、设计要求

1. 总平面图

要求：带比例尺，画出详细的总体环境布置（包括建筑、道路、广场、绿化、小品等）

2. 各层平面图及屋面排水图

3. 立面图（至少 2 个）

4. 剖面图（至少 2 个）

5. 效果图

6. 设计说明：不少于 1000 字

7. 经济技术指标：总建筑面积、用地面积、建筑占地面积、容积率、绿地率、建筑层数等

8. 必要的分析图：设计理念构思分析、功能分区、流线组织、内部空间分析等

9. 建筑单体部分要求学生运用 BIM 技术优化建筑空间设计表达

10. 汇报时可运用 PPT 汇报配合三维动画的模式，但不作强制要求

区位图

区位图

城乡规划专业毕业设计总体介绍

城乡规划专业毕业设计建设情况

1. 毕业设计选题

湖南大学城乡规划专业教学组织过程注重毕业设计。2011 年以来，逐年对毕业设计的要求做进一步强化，每年召开专题研讨会，研究命题的科学性。已逐步形成城市总体设计、特色主题城市设计、村镇规划设计、特色片区城市更新设计、"旧城栖居"联合毕设等较稳定的特色选题方向。

毕业设计选题遵循以下原则：规模在 15~20hm²（研究范围可扩大），不宜过小以免降低难度，使其复杂性不足；也不宜过大以免增加学生工作量，影响设计成果深度。毕业设计命题均以实际用地，结合实际工程要求适当调整经济技术指标，课题接近实际工程条件；毕业设计成果要求城市设计部分达到初步设计深度要求，另需提交设计前期调研报告、设计论文、规划必要图纸等，选题的内容、难度和综合性均应高于课程设计，能涵盖本科教育中的主要知识点。

2. 毕业设计指导

2008 — 2010 年，毕业设计由副教授以上职称，或工程经验丰富的教师指导完成，每位教师指导 8 名学生；2011 年之后，按照学校要求，每位教师均需参加毕业设计指导，我院因此组织了不同毕业设计小组指导毕业设计，平均每位教师指导 3 名学生；2013 年后，组织了毕业设计命题筛选活动，减少命题数量，提高命题质量，以同一命题为小组指导学生毕业设计。教学组织中，教师与学生每周见面不得少于两次，在教师指导下，由学生独立完成自己的设计任务，提交完备的毕业设计文件。在毕业设计教学组织中，强化过程控制与中期检查的重要性，避免学生对设计进度把握不足影响设计成果。

3. 毕业设计量化评分及反馈机制

毕业设计既是本科学生培养的重要环节，也是检验城乡规划教学质量的重要阶段。我院自 2012 年开始，根据教学计划中的课程组成，确定各个知识模块的权重，实施毕业设计量化评分尝试；进而应用各个知识板块的得分情况，对本科教学计划实施中的薄弱环节予以判定，形成设计教学的反馈机制，推动教学改革有目的的实施。严格控制毕业设计深度要求，严格把控毕业设计的评分过程。设计评分实施了量化评分制度，避免各个答辩小组评分不均的情况；各个答辩小组包含建筑设计、城乡规划、建筑技术专业教师，以及外聘设计院专家一名；设计成绩教师占 30%，答辩小组成绩 70%，构成合理；严格控制毕业设计成果，中期检查不合格者推迟答辩（暑期之后），答辩成绩不合格重修（来年随低年级重新修课），每年淘汰一定数量的不合格作业。

4. "旧城栖居"联合毕业设计工作营

2020 年"旧城栖居"四校"规划 + 建筑"联合毕业设计主题由湖南大学建筑与规划学院发布，联合北京工业大学、华中科技大学、西南交通大学四校举行。为青年学子、建筑学界提供了一个毕业设计的交流平台。活动采用线上方式，分享下列内容：1）毕业设计组织、命题心得，通过交流提高毕业设计命题质量；2）优秀的学生代表、毕业设计创作作品，以榜样激发学生的兴趣与创造力。此外，活动还邀请来自各大设计机构及院校的专家作为评论嘉宾，与参与分享的教师与学生展开对谈与研讨。活动至今已成功举办两届，在学界引起了较热烈的反响，活动参与人数近两万人次，为全国本学科类毕业设计注入了新的活力。

2020 年"旧城栖居"四校联合毕业设计

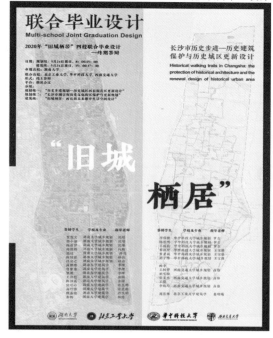

课题一

湘西怀化溆浦县城市设计

一、设计任务

溆浦县城总体城市设计 + 重点片区城市设计

二、设计目的

研究和挖掘西部地区的地域与人文特色，并将地域性考量贯穿于城市规划与景观专业的宏观、中观尺度，直至建筑专业的微观尺度。强调城市、景观、建筑三个专业的交叉与融合。

基于此宗旨，2019 年毕业设计目标是为湘西溆浦县提供具有前瞻性、整体性、地域性的城市、景观与建筑设计方案。

设计过程将划分为两阶段：

第一阶段，不同专业同学混搭分组，共同进行城市、景观和建筑调研，提出整体城市设计思路与概念性的城市片区设计方案。

第二阶段，不同专业的同学按本专业要求深化。城市规划专业进一步完善总体城市，并对重点片区进行深入城市设计，并完成城市设计图则；景观专业完善总体景观规划方案，并对重点地段进行景观设计；建筑专业着重进行文化类建筑设计。

三、主要参考资料（参数）

1.《中华人民共和国城乡规划法》

2.《规划编制方法》

3.《中华人民共和国土地管理法》

4.《溆浦县城总体规划（2005-2020）（2017 年修改）》

5.《溆浦县城其他相关规划》

6. 溆浦县城地形图 1：1000，1：10000

7.《城市规划》《建筑学报》《规划师》《理想空间》等规划杂志

溆浦县城影像图

重点片区影像图

溆浦县城位置示意

课题二

郴州市山水城市总体城市设计

一、设计目的

1. 通过本次设计课程，掌握城市设计的基本思路、流程和方法，了解城市设计任务的一般要求，完成一套较为完整的设计成果。

2. 尊重自然生态环境，追求相契合的山环水绕的形意境界，继承了中国城市发展数千年的特色和传统。建立'人工环境'（以城市为代表）与'自然环境'相融合的人类聚居环境，使人工环境与自然环境相协调发展。

3. 了解山水城市设计的一般手法和方式。

4. 进一步提高设计方案的图面和口头表达能力，熟悉设计团队运作的通常做法。

二、设计要求

1. 按照进度安排，完成各个任务环节。

2. 面对当代城市的各种现代技术、现代产业和现代社会文化特征，将中国传统精华自如地应用到现代城市规划中，还需要漫长的征途和艰辛的探索。发挥自主学习的能动性，查阅相关资料，并在此基础上提出设计思路。

3. 了解郴州市的城市发展历史，应在保存山水特色的基础上谋求城市发展的新方向。

4. 设计中努力体会并思考城市与自然的关系。

5. 学习应用相关的城市设计理论和方法。

三、主要参考资料

1. 片区已批控规（部分）

2. 城区地形图（内部保密使用）

3. 郴州市旅游发展总体规划

4.《郴州市城市总体规划 2009—2030 年》(2018 年修改）

5. 其他相关规划

郴州市城区影像图

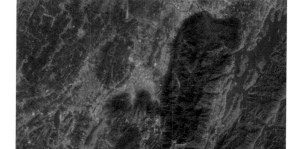

重点片区影像图

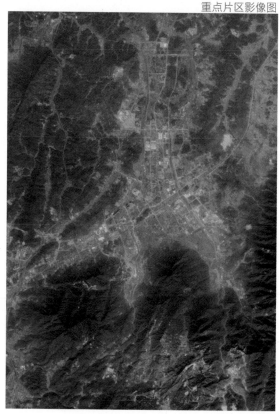

郴州市位置示意

课题三

资阳区乌龙堤村美丽乡村人居环境规划

一、设计目的

1. 通过本次设计课程，掌握村镇总体规划和村庄规划的基本思路、流程和方法，了解村镇规划任务的一般要求，完成一套较为完整的设计成果。

2. 积极应对国家战略，推动乡村事业和产业协同发展，构建和完善兜底性、普惠型、多样化的公共服务体系，不断满足乡村居民日益增长的多层次、高品质健康和发展需求。

3. 了解村镇总体规划的一般手法和方式。寻找解决城镇发展的切入点。

4. 进一步提高设计方案的图面和口头表达能力，熟悉设计团队运作的通常做法。

二、设计要求

1. 按照进度安排，完成各个任务环节。

2. 系统性的考虑：村庄规划必须以一种系统性的方式予以解决，从规划到建筑、到景观等。

3. 设计中努力体会并思考城市与建筑的关系。

4. 解决产业、公共服务等配套服务体系的配比和规划，培养处理复杂设计矛盾的基本能力。

5. 学习应用相关的城市规划和设计的理论和方法。

三、主要参考资料

1. 乌龙堤村地形图（内部保密使用）

2. 乌龙堤村相关统计年鉴

3. 乌龙堤村相关上位规划

4. 益阳市资阳区相关规划

5. 美丽乡村相关政策规划

乌龙堤村周边影像图

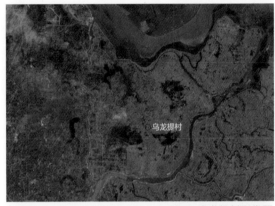

乌龙堤村影像图

乌龙堤村位置示意

课题四

岳麓区阜埠河路段及周边城市区域有机更新

一、设计目的

1. 通过本次设计课程，掌握城市更新设计的基本思路、流程和方法，了解城市设计任务的一般要求，完成一套较为完整的设计成果。

2. 了解城市更新设计的一般手法和方式。

3. 学会对客观存在实体（建筑物等硬件）改造的同时对各种生态环境、空间环境、文化环境、视觉环境、游憩环境等的改造与延续，包括邻里的社会网络结构、心理定势、情感依恋等软件的延续与更新。

4. 提高设计方案的图面和口头表达能力，熟悉设计团队运作的通常做法。

二、设计要求

1. 按照进度安排，完成各个任务环节。

2. 旧城保护与更新是世界范围内有历史城市发展面临的重要问题，发挥自主学习的能动性，查阅相关资料，并在此基础上提出设计思路。

3. 了解长沙的城市发展历史，应在保存街区特色的基础上谋求旧街区新的发展方向。

4. 设计中努力体会并思考城市与建筑的关系。

5. 学习应用相关的城市设计理论和方法。

三、主要参考资料

1. 片区已批控规（部分）

2. 片区地形图（内部保密使用）

3. 岳麓山国家大学科技城相关建设规划

4. 《长沙历史文化名城保护规划》（部分）

5. 《长沙市城市规划技术管理规定 (2018 年修订版)》

阜埠河路周边城市影像图

阜埠河路段影像图

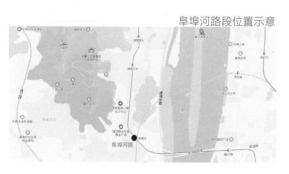

阜埠河路段位置示意

课题五

长沙市历史文化街区保护与更新规划

一、设计目的

1. 通过本次设计课程，掌握城市设计的基本思路、流程和方法，了解城市设计任务的一般要求，完成一套较为完整的设计成果。

2. 历史街区的保护与更新是城市设计任务的重要内容，通过本次设计，尝试从熟悉的城市环境出发，创造有时代感与历史意义的场所环境。

3. 了解旧城保护更新设计的一般手法和方式。

4. 进一步提高设计方案的图面和口头表达能力，熟悉设计团队运作的通常做法。

二、设计要求

1. 按照进度安排，完成各个任务环节。

2. 旧城保护与更新是世界范围内有历史城市发展面临的重要问题，发挥自主学习的能动性，查阅相关资料，并在此基础上提出设计思路。

3. 了解长沙的城市发展历史，应在保存历史特色的基础上谋求旧街区新的发展方向。

4. 设计中努力体会并思考城市与建筑的关系。

5. 历史街区具有丰富的景观与人文环境基础，同时也要考虑相关的功能等问题，培养处理复杂设计矛盾的基本能力。

6. 学习应用相关的城市设计理论和方法。

三、主要参考资料

1. 片区已批控规（部分）

2.《长沙历史文化名城保护规划》（部分）

3.《历史文化步道规划设计指引（试行）》（部分）

4. 片区地形图（内部保密使用）

5.《长沙市城市规划技术管理规定(2018年修订版)》

片区影像图

潮宗街
北正街
西长街

街区影像图

历史文化街区位置示意

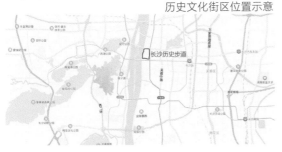

长沙历史步道

目录

专题五：公共文化建筑设计
Topic 5: Public Culture Architecture Design

专题六："旧城更新"城市与建筑设计（新四校联合毕设）
Topic 6: Urban and Architectural Design of "Old City Renewal"

城乡规划专业优秀毕业设计
专题一：湘西怀化溆浦县城市设计
Topic 1: Land Spatial Planning and Overall Urban Design

专题二：特色主题城市设计
Topic 2: Urban Design with Specific Theme

专题三：村镇规划设计
Topic 3: Village Planning and Design

专题四：城市特色片区更新规划与设计
Topic 4: Renewal Planning and Design of Urban Characteristic Areas

专题五："旧城栖居"联合毕业设计
Topic 5: Historical Trail Update Design

专题一：历史建筑虚拟修复设计
Topic 1: Virtual Restoration Design of Historical Buildings

1928 天心阁效果图

1928 天心阁鸟瞰图

设计题目： 天心阁历史资料考据与虚拟复原研究
指导老师： 钟明芳
学　　生： 高雨寒

1

Design topic: Study on Historical Data and Virtual Restoration of Tianxin Pavilion

Instructor: Zhong Mingfang

Student: Gao Yuhan

● 设计说明

天心阁始建于明代，位于长沙市旧城东南角的古城墙上，历史上屡拆屡建，见证了长沙城几百年的兴衰史。天心阁是湖南省唯一保存完整的、规模宏大的集城墙、瓮城、阁楼、园林于一体的古城池建筑群，具有极高的文化和历史价值。其中清末和1928年重建的天心阁距今较近，有着丰富的文字资料和老照片，具备较高复原价值和较好的复原条件。本次毕业设计拟通过复原风貌为长沙市历史建筑保护工作做出有益的探索和尝试；同时，学习和探索数字复原的方法，运用最新的虚拟修复的技术手段，使天心阁的面貌与细节得到更全面的展现。

Design notes

Tianxin Pavilion was built in the Ming Dynasty, located on the ancient city wall in the southeast corner of the old city of Changsha. In history, Tianxin Pavilion was repeatedly demolished and built, which witnessed the rise and fall of Changsha city for hundreds of years. Tianxin Pavilion is the only well-preserved, large-scale, set walls, wengcheng, attic, garden in one of the ancient city complex, with high cultural and historical value. The Tianxin Pavilion, which was rebuilt in the late Qing Dynasty and 1928, is relatively recent. It has a wealth of written materials and old photos, and has high restoration value and good restoration conditions.This project intends to make a beneficial exploration and attempt for the protection of historical buildings in Changsha through the restoration of features. At the same time, learning and exploring the digital restoration method, using the latest virtual restoration technology means, so that the face and details of Tianxin Pavilion get a more comprehensive display.

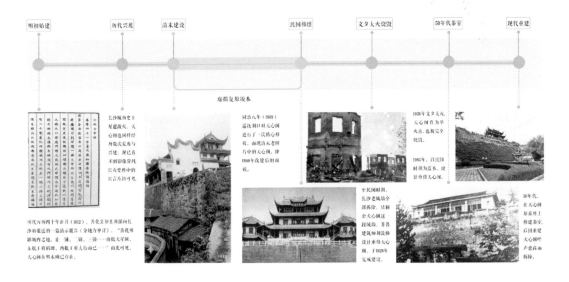

● 清末天心阁平面复原　Restored image of building plan

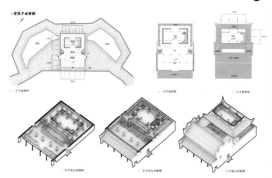

从清末天心阁的照片中可以推论，其主阁未变，在主阁前建一个两层的副阁，主副阁通过天井和两侧廊道相连；副阁前有一条通道，通道靠城墙边沿有石护栏；副阁西南和东北两端，连通天心阁东南、西南和东北三向。

It can be inferred from the photos of Tianxin Pavilion in the late Qing Dynasty that its main pavilion has not changed. In front of the main Pavilion, there is a two-story auxiliary Pavilion, which is connected with the corridors on both sides through the patio; There is a passage in front of the auxiliary Pavilion, which is surrounded by a stone fence near the wall; The southwest and northeast ends of the sub Pavilion connect the southeast, southwest and northeast directions of the Tianxin Pavilion.

● 清末天心阁形制推断　Inference of the shape

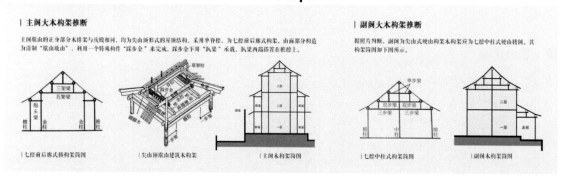

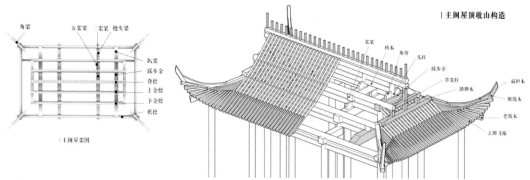

● 清末天心阁门窗推断　Inference of the doors and windows

门窗形式可以从老照片中辨认，隔扇门窗的窗棂已无法分辨，研究发现门窗装饰和岳麓书院御书楼相似，推测为湖南地区的常见做法，遂参考其门窗形式和其他装饰。

The form of doors and windows can be identified from the old photos, and the window lattice of the partition doors and windows has been unable to be distinguished. The study found that the decoration of doors and Windows is similar to that of the Yushu Building of Yuelu Academy. It is speculated that it is a common practice in Hunan province, so we refer to the form of doors and windows and other decoration.

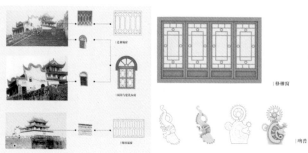

● 1928 天心阁平面复原　Restored image of building plan

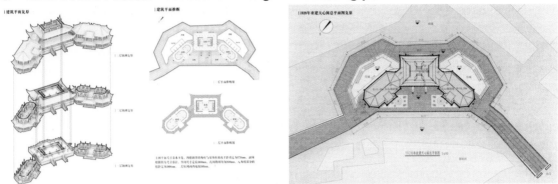

● 1928 天心阁建筑结构解析　Building structure analysis

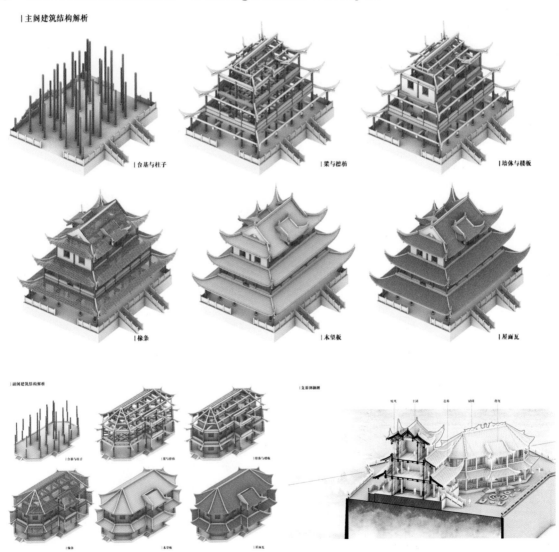

● 清末复原效果图　Restored image

清末的天心阁三面为砖墙垛墙维护，另一面通过一个天井与副阁相连。而副阁为二层硬山楼阁，其中一层部分带有前廊，照片上所显示的猫躬背垛墙即为其封火墙，与主阁围墙连接为整体。

The Tianxin Pavilion in the late Qing Dynasty was maintained by brick walls on three sides and connected to the secondary pavilion on the other side by a patio. The second pavilion is a two-story pavilion, with a front porch on the first floor. The wall shown in the photo is the firewall, which is connected with the wall of the main pavilion as a whole.

● 虚拟修复技术　Virtual restoration technology

1. 虚拟现实技术（VR）
虚拟现实（Virtual Reality）也称为虚拟技术、虚拟环境，是20世纪发展起来的一项全新的实用技术。是利用计算机模拟产生一个三维空间的虚拟世界，提供用户关于视觉等感官的模拟，让用户感觉仿佛身临其境，可以即时、没有限制地观察三维空间内内的事物。随着科技的发展，虚拟现实技术也取得了巨大进步，并逐步成为一个新的科学技术领域。

2. 增强现实技术（AR）
增强现实（Augmented Reality）是一种实时地计算摄像机影像的位置及角度并加上相应图像的技术。包含了多媒体、三维建模、实时视频显示及控制、多传感器融合、实时跟踪及注册、场景融合等新技术与新手段。它将计算机生成的虚拟物体或关于真实物体的非见们信息叠加到自实世界的场景之上，实现了对真实世界的增强。

3. 全息投影技术
全息投影技术（Front projected Holographic Display）属于3D技术的一种。原像利用干涉和衍射原理记录并再现物体真实的三维图像的技术。利用目视效和光学原理，制造出真实虚度空间内的立体影像。

4. 动画复原技术
通过三维建模和场景动画制作，得动画投射到大屏幕上，使人快速直观地看到历史建筑的动态与效果。这是历史建筑复原常用的技术手段之一，也是本次设计的重点研究内容。

目前在手机的相应程序上即可支持 AR 技术，设想可以采集现在天心阁的相关影像信息，通过网络上现有的微信 AR 小程序进行相关制作，使人们使用手机、照相机对准天心阁的场景，就可以看到历史上天心阁的立体影像。

At present, AR technology can be supported on the corresponding program of the mobile phone, and it is conceived that the relevant image information of the current Tianxin Pavilion can be collected, and the relevant production can be made through the existing WeChat AR mini program on the network, so that people can use mobile phones and cameras to point at the scene of Tianxin Pavilion, and they can see the stereoscopic images of Tianxin Pavilion in history.

● 更新利用　Renewal and utilization

| 天心阁一层展览

| 天心阁二层展览

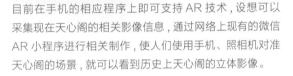

现存天心阁的主阁仍为三层，其中一层、二层已有展览，只有三层闲置，则三层可以作为历史上两个版本的天心阁虚拟展示的地方。主要展示的形式为，在中间放置两个 3D 全息投影展柜，进行模型展示；在两侧悬挂幕布，分别播放虚拟复原动画。

The main pavilion of the existing Tianxin Pavilion still has three floors, among which the first and second floors have been exhibited, and only the third floor is idle, so the third floor can be used as a virtual display place for the two versions of Tianxin Pavilion in history. The main forms of display are: two 3D holographic projection display cases are placed in the middle for model display; hang curtains on both sides and play virtual restoration animations.

根据图纸修复效果图

室内修复效果图

设计题目： 湖南大学图书馆旧址虚拟修复与系统更新设计
指导老师： 钟明芳
学　　生： 秦浦斯

2

Design topic: Virtual Restoration and System Update Design of the Old Site of Hunan University Library

Instructor: Zhong Mingfang

Student: Qin Pusi

● **设计说明**

近年来随着科技的进步以及人们对于历史建筑的重视，使得虚拟修复历史建筑的理论与实践得到蓬勃发展。本次课题的研究对象是一座几乎完全损毁消失的建筑，其遗留物不足整体的1%，仅几根残留的柱子。但是此次虚拟修复仍然坚持贯彻《威尼斯宪章》的指导原则，对于历史建筑的原真性给予充分的尊重与坚持，不随意臆测，修复后仍将其开放，发挥它的作用。

在此基础上，本次虚拟修复及其更新系统设计，还将创新地使用多种现代科学技术与创新虚拟修复的理念，通过加建、增建、功能恢复等手法来使历史建筑"复活"。

Design notes

In recent years, with the progress of science and technology, and the importance attached to historical buildings, the theory and practice of virtual restoration of historical buildings are booming. The research object of this project is an almost completely destroyed and disappeared building, with less than 1% of the total remains, only a few remaining columns. However, this virtual restoration still adheres to the guiding principles of the Venice Charter, fully respects and insists on the authenticity of historical buildings, and does not speculate at will. After restoration, it is still open to play its role.

On this basis, the virtual restoration and its update system design will also make innovative use of a variety of modern technology to innovate the concept of virtual restoration, and "revive" historical buildings by means of addition and functional restoration.

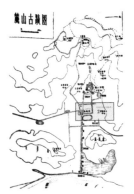

● 调研测绘　Survey and mapping

针对残存的几根花岗石柱的样式、细节、纹饰等现场进行测绘分析，采集数据做法样本。特别关注了柱子上抗战时期留下的弹孔，这是历史建筑作为史料最重要的特点。另外，对同时期同建筑师的作品进行了对比和调研分析，特别是距离图书馆很近的工程馆，具有较高的参考价值。

In view of the style, details and ornamentation of several remaining granite columns, surveying and mapping analysis was carried out on site, and data samples were collected. Special attention was paid to the bullet hole left on the column during the Anti-Japanese War, which is the most important characteristic of historical buildings as historical materials. In addition, compared with the works of the same architects in the same period, the research and analysis, especially the engineering building, which is very close to the library and has high reference value.

● 史料分析　Historical data analysis

湖南大学原图书馆是 1929 年 4 月由任凯南校长请省政府拨款兴建的，同年 11 月由蔡泽奉教授设计，1930 年春动工，1933 年 9 月在胡庶华校长任内落成，耗资 88370 元。

湖南大学原图书馆于 1938 年毁于日本帝国主义的野蛮轰炸之中。1938 年 4 月 10 日，日本侵略军的飞机 3 次侵犯长沙领空，每次 9 架飞机，对湖南大学等校区实施狂轰滥炸，投掷炸弹、燃烧弹 200 余枚，湖南大学顿时陷入火海，图书馆变成一片废墟，馆内 20 余万册图书和大批珍品毁于一旦。

The original library of Hunan University was built in April 1929 at the request of President Ren Kainan and funded by the provincial government. It was designed by Professor Cai Zefeng in November of the same year. Construction began in the spring of 1930 and was completed in September 1933 under president Hu Shuhua's presidency, costing 88,370 yuan.

The original library of Hunan University was destroyed by the brutal bombing of Japanese imperialism in 1938. On April 10, 1938, planes of the Japanese army invaded the airspace of Changsha for three times, with nine planes each time, and bombarded Hunan University and other campuses with bombs and more than 200 incendiary bombs. Hunan University was suddenly engulfed in flames, and the library was turned into ruins, with more than 200,000 books and a large number of treasures destroyed.

● 图纸修复　Drawing repair

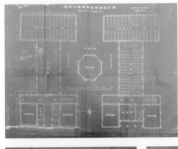
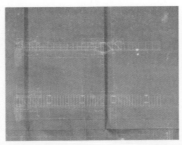

从湖南大学图书馆档案馆调取到湖南大学原图书馆相关图纸资料，对图书馆图纸部分进行修复。资料收集，是历史建筑研究的重要方法。

The relevant drawing materials of the original library of Hunan University were retrieved from the archives center of Hunan University library, and some drawings of the library were repaired. Data collection is an important method of historical building research.

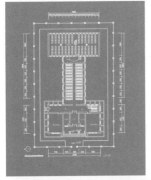

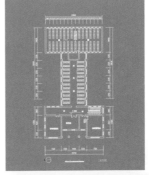

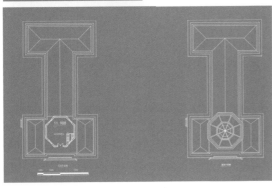
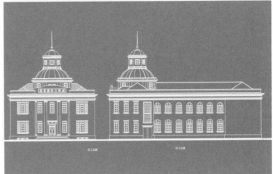

● 结构分析　Structure analysis

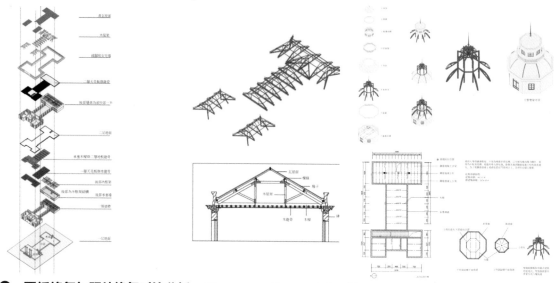

● 图纸修复与照片修复对比分析　Comparative analysis

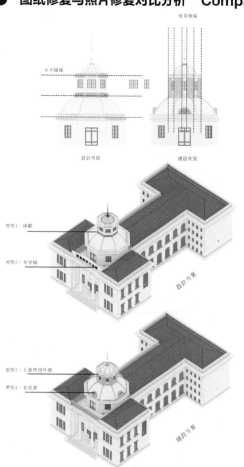

差别 1、3：分析可知在实际施工建设中，为了突出穹顶的统领地位，将原来的线脚取消，增加了侧墙的竖向线条，使得窗户比例变长，突出垂直向，视觉上拉长穹顶高度。

差别 2：推测由于建设年代历史背景原因，社会动荡不安，为了避免敌军突袭而有意去掉石碑。

差别 4：推测建成后为了改善通风采光，或者之前观察视野不佳，从而增加老虎窗便于观察。

根据图纸修复

根据照片修复

● 虚拟技术应用　Application of virtual technology

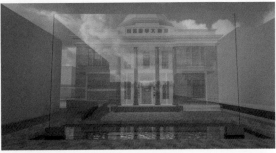

高清投影膜展示效果图（设备开启前）　　　　高清投影膜展示效果图（设备开启后）

AR 虚拟修复效果图　　　　360 度全息投影展示柜效果图　　　　VR 虚拟修复效果图

● 更新利用　Renewal and utilization

分别在原来的草坪上置入三个建筑体量，通过抬升架空，使用中性材料，利用可逆性的结构施工技术，以一种低姿态介入这片近代历史建筑群。然后通过连廊、入口广场将四栋建筑串联起来，形成一个完整的展览序列。通过拉长流线，并利用玻璃的通透性，在行径中参观实现与周边环境的融入，实现与周边的近代历史建筑互动对话。

Three building volumes are placed on the original lawn respectively. By lifting overhead, using neutral materials and reversibility structural construction technology, the modern historical architectural complex is intervened in a low posture. Then the four buildings are connected through the corridor and the entrance square to form a complete exhibition sequence. By elongating the streamline and taking advantage of the permeability of glass, the tour realizes the integration with the surrounding environment, realizes the integration with the surrounding area, and realizes the interactive dialogue with the surrounding modern historical buildings.

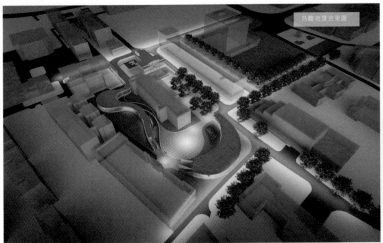

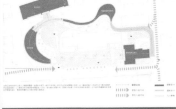

专题二：绿色高层建筑设计
Topic 2: Green High-rise Building Design

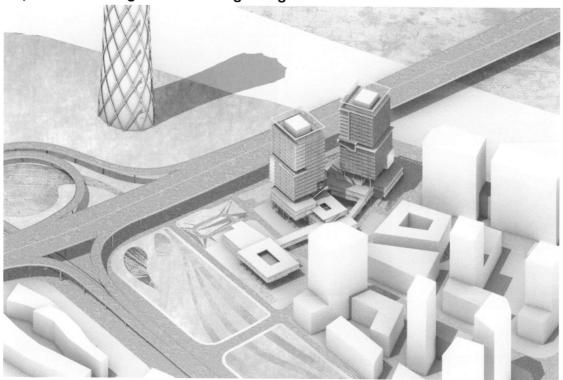

鸟瞰图

局部效果图

设计题目：湖南创意社区高层建筑设计
指导老师：谢菲
学　　生：刘航

3

Design topic: High-rise Building Design of Hunan Creative Community

Instructor: Xie Fei

Student: Liu Hang

● **设计说明**

项目基地选址处于长沙视频文创产业园内。产业园处于长沙城市升级发展的重要节点上，建设生态绿色的新型文化园，既要服务于建设基地周边现有的生活，拉动区域活力，同时也要提升湘江片区，造就"新东方莱茵河"地标性的城市生态绿色空间。设计项目空间规划充分考察设计场地及其周边环境的时空特点，反映所处的地域气候，探讨适宜现代城市人文和社会价值观的公共文化空间形态或设计模式，这正是本设计项目开展的初衷。设计需要结合场所未来的空间演化进行诠释和分析，设想和策划适宜基地及其环境的新兴的文创服务设施和建筑空间。

Design notes

The project base is located in Changsha Video Cultural and Creative Industrial Park. The industrial park is located at the key node of changsha's urban upgrading and development. The construction of a new ecological and green cultural park should not only serve the existing life-driving regional vitality around the construction base, but also improve the Xiangjiang Area and create a landmark urban ecological and green space of "New Oriental Rhine". The original intention of the design project is to fully investigate the spatial and temporal characteristics of the design site and its surrounding environment and reflect the regional climate, and to explore the public cultural space form or design mode suitable for modern urban humanistic and social values. The design needs to interpret and analyze the future spatial evolution of the site, envisage and plan the emerging cultural and creative service facilities and architectural space suitable for the site and its environment.

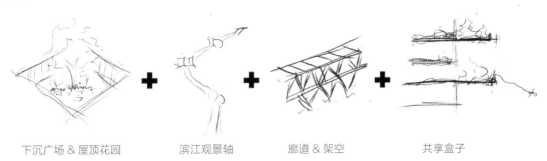

下沉广场 & 屋顶花园　　　　滨江观景轴　　　　廊道 & 架空　　　　共享盒子

● 形体生成　Shape generation

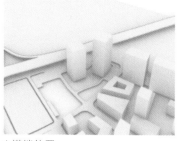 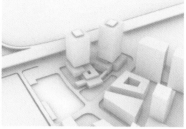 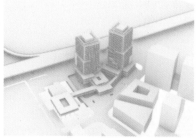

1 塔楼位置　　　　　　2 裙楼适应周边环境　　　　　3 置入盒子 + 绿化 & 表皮植入

● 场地规划　Site planning

● 绿建策略　Green building strategy

设计类	平面布置 控制体型系数，体型系数越大、外表面积较大，热量越容易流失
节水与水资源利用	雨水回收系统（雨水充足适合使用） 绿化节水灌溉（雨水多不需要） 中水利用（寒冷地区雨水少适合，这里不必）
节地与室外环境	场地生态保持 屋顶绿化 地面铺装 地下空间利用
节能与能源利用	围护结构保温、建筑外遮阳、屋顶太阳光系统等形式 空调系统 照明系统
室内环境	自然通风 建筑外遮阳——横竖构件遮阳、挑板阳台、体块自遮阳 阳光车库
隔声系统	加强围护构件，特别是幕墙的隔声性能（Low-E玻璃和中空玻璃） 根据场地风环境和声环境合理设计开窗位置 充分利用绿化与遮挡的降噪效果（竖向可调节遮阳百叶、网架式垂直绿化） 精心选择幕墙通风器安装位置，兼顾通风与隔声
运营管理类	智能化系统运营

在建筑中部加强防噪声措施，例如竖向可调节遮阳百叶、网架式垂直绿化。如果高层住宅周围存在真空状态，即没有绿化或建筑物，则噪声的传播应遵循从下到上逐渐减小的规律。然而，在实际环境中，由于各种建筑物的布置和绿化措施，噪声具有"中间大而两端小"的布局。

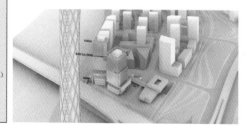

初步模拟 · 风环境 & 声环境　Preliminary simulation

初步模拟-风环境

设置标准：夏季风速2.8m/s，冬季风速2.4m/s，高度H=1.5m

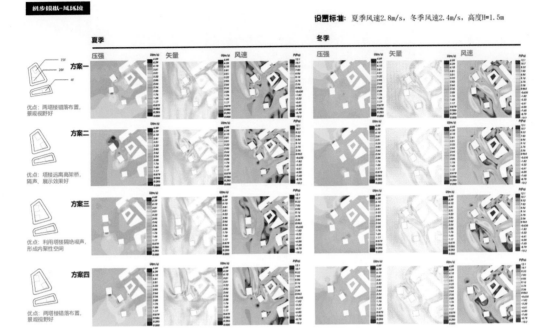

初步模拟-声环境

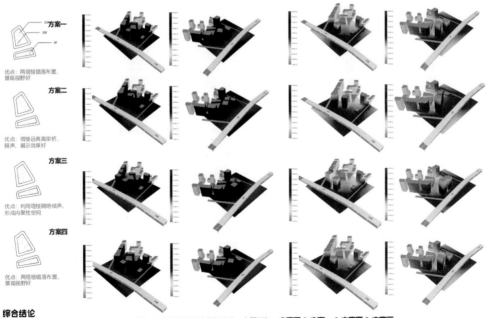

综合结论

已知场地的主要噪声来源于西侧，因此主要针对西侧立面的影响进行讨论。由图可知，**方案三 > 方案一 > 方案四 > 方案二**综合声环境，以及结合对于浏阳河的景观视线问题选取**方案一**进行深化。

● 深化模拟·风环境优化　Deepening simulation

冬季-原本

冬季-18m处A平台

冬季-25m处B平台
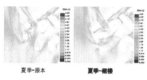
夏季-原本　　　　夏季-裙楼

风速度云图——加入导风盒子（左图红色示）后A平台处的静风区面积明显减少，同时南侧建筑东北角的风速同样减缓B平台处的静风区同样减少，比较适宜

速度矢量图——设置裙楼之后，两塔楼之间的涡旋消失，可以作为人的活动区域

问题：
　　冬季南部塔楼有大量静风区，夏季两塔楼之间有涡卷。
解决方法：
　　导风盒子。

Problem:
　　There are a large number of calm wind areas in the Southern Tower in winter and vortices between the two towers in summer.

Solution:
Air guide box.

● 被动式遮阳设计　Passive sunshade design

Incident Radiation

1) 水平遮阳模拟

未加装构件总辐射量

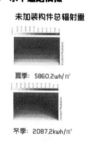
夏季：5860.2wh/m²

冬季：2087.2kwh/m²

夏季遮阳率　冬季遮阳率　夏-冬遮阳率差值

• **结果分析**
由图表可知，随着水平遮阳板长度增加夏季和冬季的遮阳率逐渐增大，差值基本稳定，考虑到夏季得热少和冬季得热多的需求，确定水平遮阳板长度为1m。

2) 竖向遮阳模拟

遮阳方式：考虑到形式，采用综合式遮阳

以B栋办公楼西侧为例，用模拟软件分析自然采光系数和夏季辐射量，作为对照。

Incident Radiation

Incident Radiation

构件模拟变量条件
考虑形式

板长： 500mm、600mm、700mm、800mm、900mm

间距： 800mm、900mm、1000mm、1100mm、1200mm、1300mm、1400mm、1500mm

角度： -45°、-30°、-15°、0°、15°、30°、45°
（西偏南为正）

由图表可知，考虑到夏季得热少和冬季得热多的需求，确定水平遮阳板长度为1m。
遮阳方式：考虑到形式，采用综合式遮阳，用模拟软件分析自然采光系数和夏季辐射量，作为对照。

As can be seen from the chart, considering the demand of gaining less heat in summer and more heat in winter, the length of horizontal sunshade was determined to be 1m.
Shading method: Considering the form, comprehensive shading is adopted, and the simulation software is used to analyze the natural lighting coefficient and the amount of radiation in summer as the control.

● 被动式遮阳设计 · 竖向遮阳模拟　　**Vertical shading simulation**

构建模拟变量条件

板长：500mm、600mm、700mm、800mm、900mm

间距：800mm、900mm、1000mm、1100mm、1200mm、1300mm、1400mm、1500mm

角度：－45°、－30°、－15°、0°、15°、30°、45°（西偏南为正）

- 固定条件：板长500mm。
- 变量：间距800mm——1500mm、角度－45°——45°

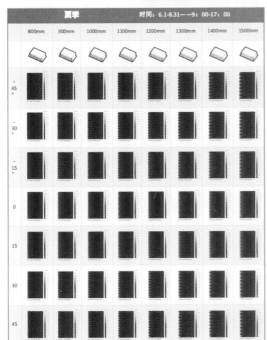

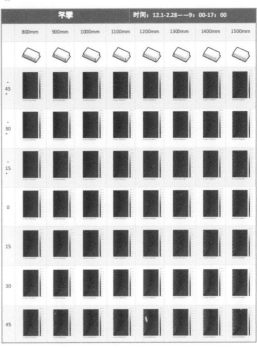

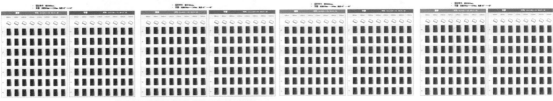

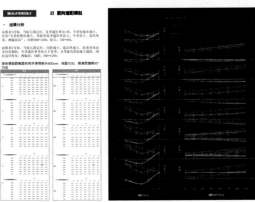

结果分析

综合遮阳系数 S_W= 外窗遮阳系数 S_C × 外遮阳系数 SD=0.88 × 玻璃遮阳系数 S_e × 外遮阳系数 SD。即为了满足规范 0.68 × 0.8 × x ≤ 0.52，得 x ≤ 0.96，即采取遮阳系数 0.96 以下的玻璃即可得出：板长 800mm、间距 1200mm、角度西偏南 30°。

Comprehensive shading coefficient SW = external window shading coefficient SC X external shading coefficient SD = 0.88 X glass shading coefficient se X external shading coefficient SD. That is, in order to meet the specification of 0.68 X 0.8 X x ≤ 0.52, x ≤ 0.96 is obtained, that is, the glass with shading coefficient less than 0.96 can be obtained: the plate length is 800mm, the spacing is 1200, and the angle is 30 ° West by south.

整体效果图

局部效果图

4

设计题目： 湖南创意文化设计总部大厦
指导老师： 刘宏成
学　　生： 徐昕

Design topic: The Design of Headquarters Building of Hunan Creative Community
Instructor: Liu Hongcheng
Student: Xu Xin

● **设计说明**

场地位于马栏山创智园片区，因此具有多元丰富的媒体资源；它同时被浏阳河环绕，自然景观优美，高架桥、计划修建的空轨和地铁站使其交通便捷，具有发展潜力。本项目的主体建筑为两栋商业办公相结合的综合体大厦，服务于新媒体工作者、创客、研发工作者以及外来游客。以四个"桥梁"（健康桥、社区桥、文化桥和培养桥）为媒介打造富有活力的办公空间，室外的公共平台与屋顶花园创造了人与人交往的可能。同时阶梯式绿化休息空间也提供了优美的景观和适宜的微气候。室内跑道和健身房也为使用者提供了更为健康的生活方式。同时以绿色建筑为目标，对场地的风环境、噪声以及采光做出分析与优化设计。

Design notes

The site is located in Malan Mountain Wisdom, so it has diversified and rich media resources. At the same time, it is surrounded by Liuyang River and has beautiful natural landscape. Viaduct, planned air track and subway station make transportation convenient and has development potential. The main buildings of the project are two complex buildings combining business and office, serving new media workers, makers, researchers and foreign tourists. Four "Bridges" (Health bridge, Community bridge, Cultural bridge and Nurturing bridge) are used to create a vibrant office space. Outdoor public platforms and roof gardens create the possibility of human interaction. At the same time, the stepped green rest space also provides beautiful landscape and suitable micro climate. Indoor running tracks and gyms also offer users a healthier lifestyle. At the same time, the wind environment, noise and lighting of the site are analyzed and optimized for the goal of green building.

● 基地分析　Base analysis

两块规划地块为商业办公用地，在马栏山创智园片区内，服务于媒体工作者和其他城市居民。计划修建的空轨和地铁、对内的公交车线路、滨河的自行车道和片区内的漫步走道便捷了工作人员与外来游客的通勤与观光。

The two planned plots are used for commercial offices, serving media workers and other urban residents in the Malan Mountain Wisdom Park area. The planned skyrail and subway, intra-city bus lines, riverside bike paths and walkways within the area facilitate commuting and sightseeing for staff and visitors.

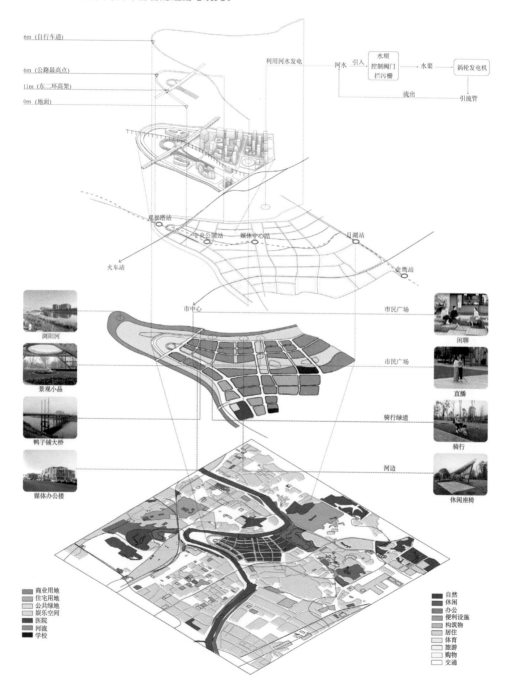

未来预期　Future expectations

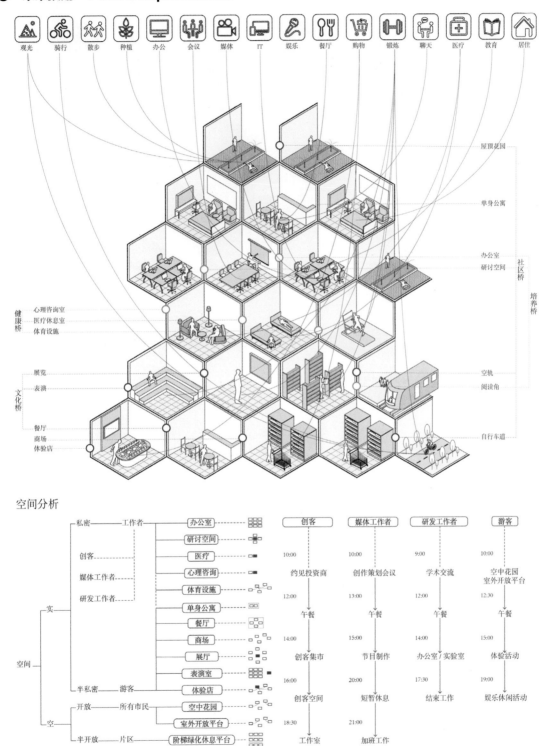

空间分析

23

气象数据　Meteorological data

气象数据的分析内容分为干球温度，相对湿度，风环境，太阳辐射和日照小时数。

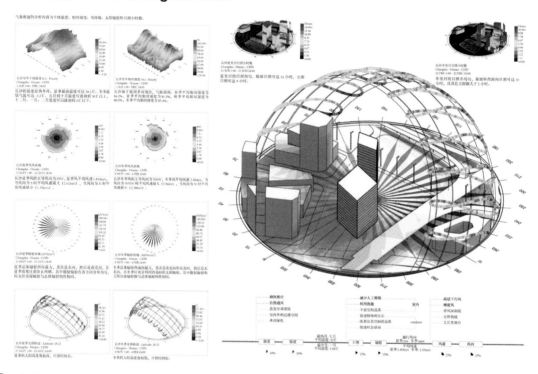

绿色建筑策略　Green building strategy

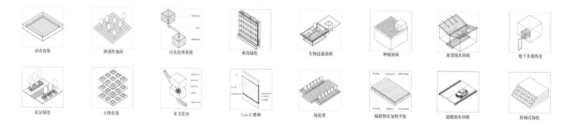

绿色建筑专题·室外噪声 & 室内采光　Special topic of green building

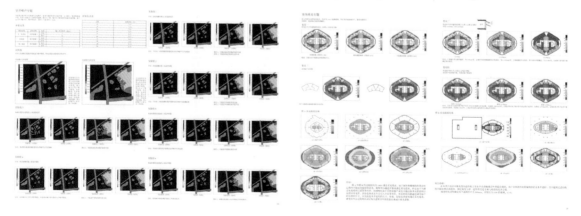

● 绿色建筑专题　Green Building

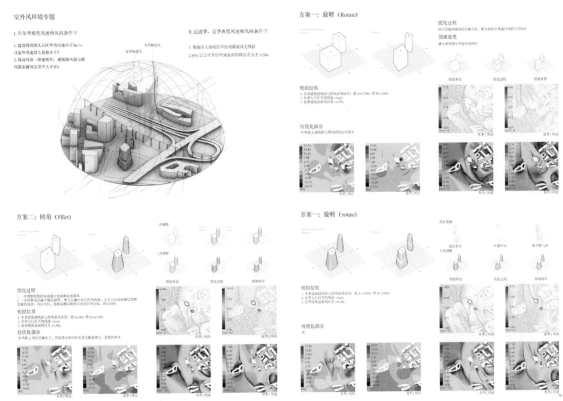

室外风环境专题

1. 在冬季典型风速和风向条件下

1. 建筑物周围人行区平均风速小于5m/s，
且室外风速放大系数小于2
2. 除迎风第一排建筑外，建筑迎风面与背
风面表面风压差不大于5Pa

II. 过渡季、夏季典型风速和风向条件下

1. 塔地库人活动区不出现滞风或无风区
2. 50%以上可开启外窗表面的风压差大于0.5Pa

方案一：旋转（Rotate）

优化过程

预期效果

模拟结果

待优化部分

方案二：圆角（Fillet）

优化过程

模拟结果

待优化部分

方案一：旋转（rotate）

优化策略

模拟结果

待优化部分

● 功能分区　Functional partition

塔楼 A 内部功能为办公室、商场、公寓、室内跑道、健身房、室外开放平台、屋顶花园、地下车库。塔楼 B 内部功能为办公室、公寓、阶梯绿化平台、室外开放平台、屋顶花园、地下车库。

The internal functions of tower a are office, shopping mall, apartment, indoor runway, gym, outdoor open platform, roof garden and underground garage. The internal functions of Tower B are office, apartment, step greening platform, outdoor open platform, roof garden and underground garage.

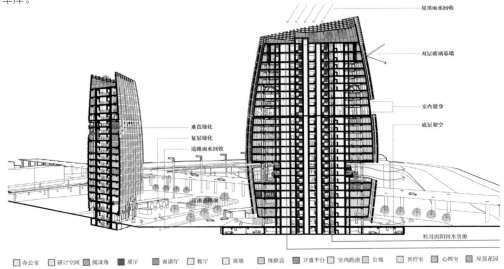

专题三：养老设施建筑设计
Topic 3: Architectural Design of Elderly Care Facilities

室外效果图

室内效果图

5

设计题目：归家之家——疗养院设计
指导老师： 张蔚 何照明
学　　生： 易紫薇

Design topic: Feeling at Home —— Sanatorium Design
Instructors: Zhang Wei, He Zhaoming
Student: Yi Ziwei

● 设计说明

二里牌社区疗养院中心位于湖南省长沙市芙蓉区二里牌社区内部。紧邻华天酒店北侧，位于多个老年社区交接的位置，该项目拟建一栋建筑面积约为 12000m²、建筑高度不超过 24m 的医养结合的社区疗养中心。本设计依托医养结合模式、就地养老政策，为社区病后失能康复老人、失忆失智老人、孤寡老人营造康复治疗、护理关怀的康复治疗中心；尽量消解机构化养老的各种弊端，依据多元化的治疗和护理模式为老年人营造多义康复养老生活空间——一个温馨的"归家之家"。

Design notes

Erlipai Community Sanatorium Center is located in Erlipai Community, Furong District, Changsha City, Hunan Province. Adjacent to the north side of Huatian Hotel and located at the junction of several elderly communities, the project plans to build a community nursing center covering an area of about 12,000 square meters with the height of no more than 24 meters. This design relies on the mode of combining medical care and nursing care, and the local pension policy, to build a rehabilitation treatment and nursing care center for the elderly with disability rehabilitation, amnesia and dementia, and the elderly alone. Try to eliminate all kinds of disadvantages of institutionalized endowment, and create a multi-meaning rehabilitation endowment living space for the elderly—a warm "Sanatorium line feeling at home" based on diversified treatment and nursing modes.

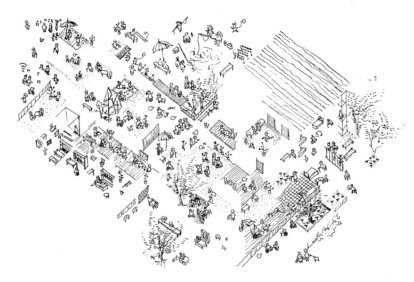

● 地块分析　Plot analysis

■ 社区空间需求分析

二里牌社区公共空间极少，只有一个小花园为主要公共空间

社区公共小花园拟建基地所在地点

■ 基地整体定位：将整个建筑处理成多个绿地空间组合形式

'green' not 'buildings'

■ 基地交通分析

- - - 外围城市车道
———— 社区内部道路

● 调研分析　Investigation analysis

红枫康乐养老院　公共活动空间

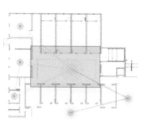

基督教城北塘养老院

长沙市第一福利院

走廊式交往空间
公共交往空间
室内封闭交往空间
交往空间

空间布局
居住
公共
辅助

需：老人在客厅、餐厅、聚会堂等公共居室空间逗留时间约为大半天，即使是独居的老人，也都喜欢打开临走廊的门窗，在室内干自己的事。老人的常驻空间为居室等公共空间。

供：现有居室空间大多为单廊走道或者临近走道的一个稍放大的空间，采光不好，空间形式单一，老人活动受限制，而老人活动是自由式的、不确定性较大，其活动空间不宜太规则，而应该是有更多的可能的多义空间，引入光、声、风，使不确定性更加自然。

Needs: The elderly stay in the living room, dining room, party hall and other public room space for about half a day, even the elderly alone like to open the door and window which adjacent to the corridor to do their own things in the room. The resident space for the elderly are the living room and other public space.

For: Most of the existing living space is a single corridor or a magnifying space near the corridor, with poor lighting and single space form, which limits the activities of the elderly. But the senior activities are freestyle. Uncertainty, its activity space should not be too regular but a polysemous space, the introduction of light, sound and wind, makes uncertainty more natural.

● 心理分析　psychoanalysis

从老年人心理活动分析/调研方式/访谈
三家现有疗养院老人走访资料

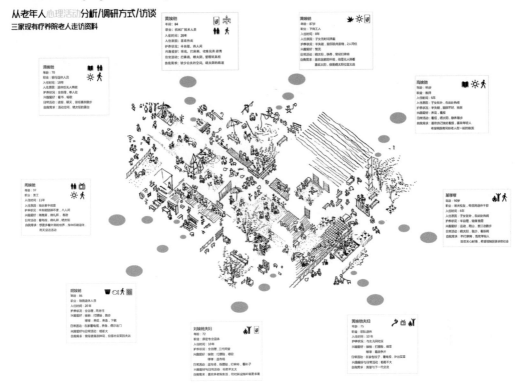

● 形体生成　Shape generation

通过前期调研，发现老人的行为特征表现为兴趣爱好很多，但是受限于腿脚不利索，所以在疗养院空间模型探索中，我们将常驻行为空间与爱好相结合，形成开敞的流动的居室空间，亦园亦家，亦公亦私。

Through the preliminary investigation, we found that the behavior of the elderly is characterized by many interests and hobbies, but their behaviors are limited by their weak legs and feet. Therefore, in the exploration of the nursing home space model, we combined the resident behavior space with hobbies to form an open and flowing living space, which is both a park and a home, public and private.

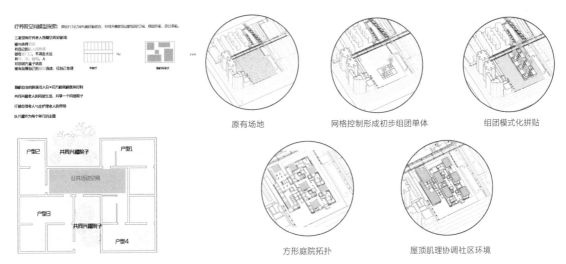

原有场地　　网格控制形成初步组团单体　　组团模式化拼贴

方形庭院拓扑　　屋顶肌理协调社区环境

● 建筑与场地多层空间关系　Spatial relationship

设计分为两个大的功能区：医疗中心和养老中心。

医疗中心 2 层，分为门诊与治疗部；养老中心 1 层为开放空间兼具日间照料、护理之用，2~4 层为疗养的居室。

The design is divided into two large functional areas: the medical center and the pension center.Medical center has 2 floors, which is divided into outpatient and treatment department; The 1st floor of the pension center is an open space for day care and nursing, while the 2nd to 4th floors are convalescent rooms.

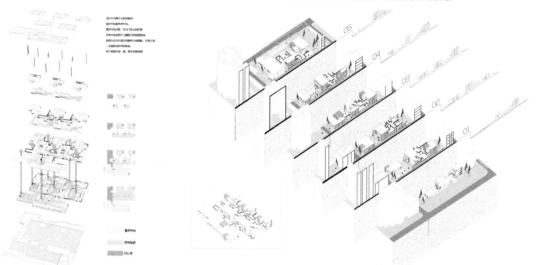

● 技术图纸　Technical drawings

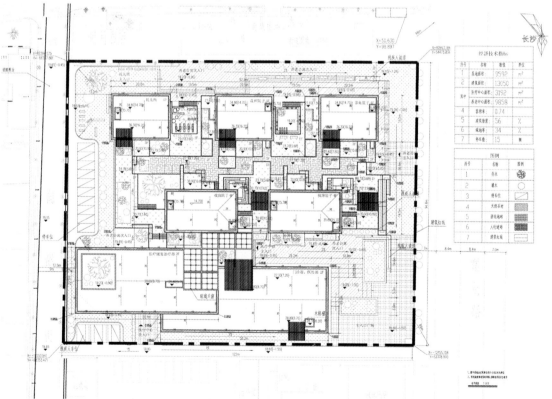

● **情景演绎　Scene interpretation**

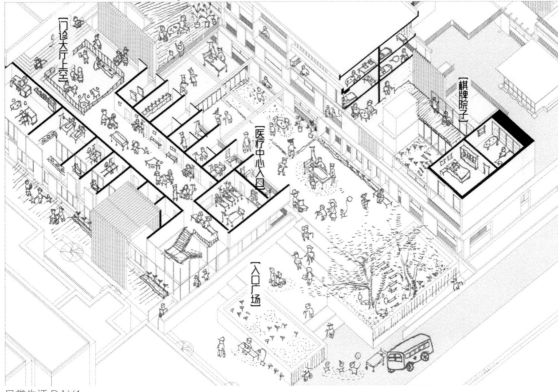

日常生活 DAY1

日常生活 DAY2

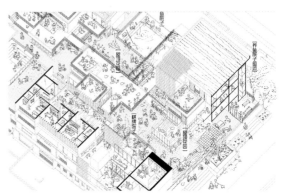

日常生活 DAY3

本项目区地处经济繁华地带，辖区内有向韶街、明月街两条主要街道，是多个社区的一个交接点，服务于整片社区。建成后除了作为该片区最大的医疗中心，还将成为这片社区居民们的一个日常生活的休憩地，并非作为医院，更多的会兼容一定的社区活动中心的功能。绿化面积并不多，因此在设计的时候将更多地给社区居民开放绿地。项目将自然融入设计，为建筑创造良好的自然环境，为疗养的居民创造良好的景观品质。

This project area is located in the flourishing economic zone, this district has two main streets: the Xiangshao street and the bright moon street, is one of the several communities intersection, serves the whole community. After the completion of the project, in addition to being the largest medical center in the area, it will also become a rest place for the daily life of the community residents. Instead of being a hospital, it will largely be compatible with the functions of a central community activity center. Generally, the green area is not much, so in the design, there is more open green space for community residents. The project integrates nature into the design to create a good natural environment for the building and a good landscape quality for the residents.

教堂视角

主入口透视

6

设计题目： 平行空间的记忆 —— 重构老人离家之"家"
指导老师： 蒋甦琦 邓广
学　　生： 沈洁

Design topic: Memory in Parallel Space —— Reconstruct the "Home" of the Old People Leaving Home

Instructors: Jiang Suqi, Deng Guang

Student: Shen Jie

● 设计说明

项目主要利用城北堂现存养老院的整个边界线，在原来的基地上重新建造一个适合老人居住的生活空间。基地东面是一片属于教堂的地，是一个废弃的女子高中，有一个较大的废弃操场，考虑将废弃操场作为整个场地的景观预留用地；在基地南侧靠近湘春巷和城北堂交界入口设置一个小型公共广场，可以吸引人流走进基地与老人多交流，同时也为基地中的教堂提供一个活动场地。基地底层为比较开放的公共活动场地，社区的人流可以进入其中与老人交流，同时底层设置多个绿化庭院，增加场地绿化面积的同时，为社区人群提供一个互动的场地。

Design notes

The project mainly makes use of whole boundary of existing nursing homes to rebuilt a living space for the elder in the original base. The eastern part of the base belongs to the church, which is an abandoned woman high school with a large abandoned playground considered as the site of landscape reserve land. A small public square is set in the south of the base near the junction of Xiangchun Lane and Chengbei Hall, which can attract people to enter the base to communicate with the elderly, and at the same time provide an activity site for the church in the base. The ground floor of the base is a relatively open public activity site, in which people from the community can enter to communicate with the elderly. At the same time, multiple green courtyards are set on the ground floor to increase the green area of the site and provide an interactive site for community people.

■ 现存老人院位置

■■ 教堂位置

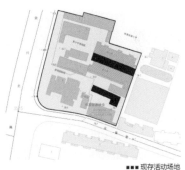

■■■ 现存活动场地

● 前期调研　Preliminary research

城北堂 人员调研

● 平面生成　Plane generation

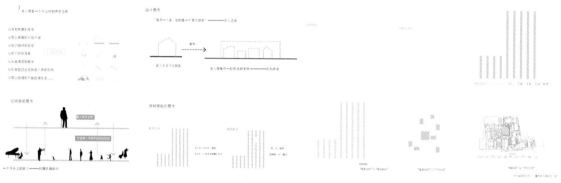

● 设计策略　Design strategy

老人组团居住空间主要以兴趣爱好分类，组团居住的老人相对活动能力较强。

Group living space for the elderly is mainly classified by interests and hobbies. The elderly who live in group have relatively strong activity ability.

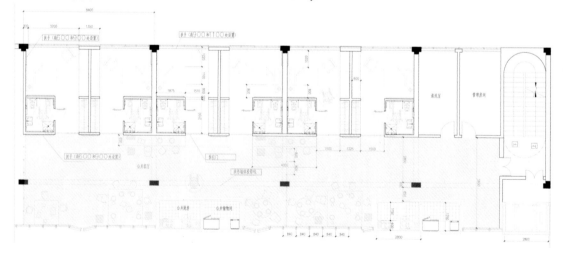

● 居住组团空间分析　Cluster spatial analysis

二层以上主要为老人居住单元空间；居住单元中置入小庭院，让老人在日常生活中可以穿越一个个小院子，获得乐趣；同时老人居住组团主要按兴趣爱好分类，让有相同爱好的老人能够有更多的交流机会。

Above the second floor are mainly residential units for the elderly; Small courtyards are placed in the living unit, so that the elderly can pass through small courtyards and have fun in their daily life; At the same time, the elderly living groups are mainly classified according to their interests and hobbies, so that the elderly with the same hobbies can have more communication opportunities.

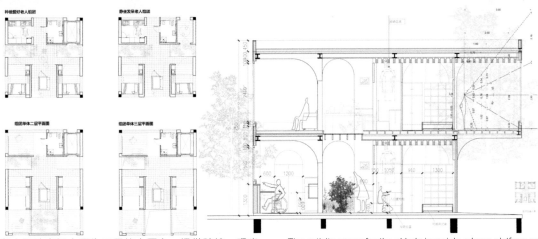

老人活动空间主要为二层的大平台，提供种植、观坐、麻将等活动，在二层有充分的交流，同时也可以直达一层公共活动区域参与活动。底层为比较开放的公共社区活动空间，社区的老人以及小孩、外来人员等不同年龄层次的人都可以来到场地进行活动，老人可以透过二层的庭院观察到底层的活动，从而便于有意愿的老人加入其中参与活动。

The activity space for the elderly is mainly a large platform on the second floor, which provides the elderly with activities such as planting, sitting and mahjong. There are sufficient exchanges on the second floor, and they can also directly participate in activities in the public activity area on the first floor. The ground floor is a relatively open space for public community activities. The elderly, children and other people of different ages in the community can come to the site for activities. The elderly can observe the activities on the ground floor through the courtyard on the second floor, so that they can participate in the activities if they are willing.

● 结构体系与功能分析　Structural system and functional analysis

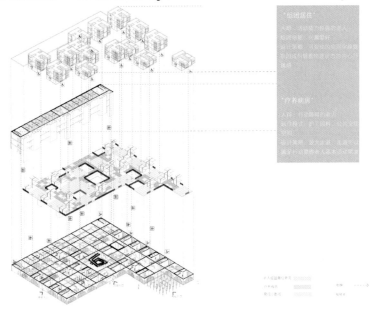

● 总平面图　General layout

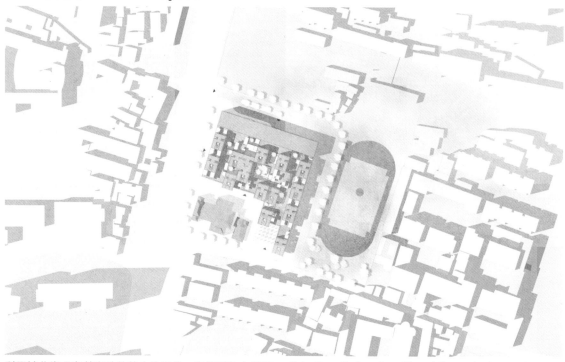

利用城北堂现存养老院的整个边界线，在原基地上重建一个适合老人居住的生活空间，基地东面作为整个场地的景观预留用地；在基地南侧靠近湘春巷和城北堂交界入口设置一个小型公共广场，可以吸引人流进入基地与老人多交流。底层设置多个绿化庭院，增加场地的绿化面积的同时为老人和社区人群的交流提供一个互动的场地。

Using the entire boundary of the existing nursing home in Chengbei Hall, a suitable living space for the elderly is rebuilt on the original site, and the east of the site is reserved for the landscape of the whole site. A small public square is set in the south of the base near the junction of Xiangchun Lane and Chengbei Hall, which can attract people to enter the base and communicate with the elderly. Several green courtyards are set up at the ground floor, which increase the green space and provide a interactive platform for the communication between the elderly and people in the community.

● 模型展示　Model show

● 效果图　Rendering

底层中央内院活动透视

主入口透视

教堂视廊

疗养病房楼"放大走廊"透视

二层街道空间

湘春巷入口空间

夏

局部效果图

鸟瞰图

7

设计题目： 洋湖老人疗养康复中心设计
指导老师： 张蔚
学　　生： 杨秾全

Design topic: Design of Yanghu Elderly Recuperation and Rehabilitation Center

Instructor: Zhang Wei

Student: Yang Nongquan

● **设计说明**

为所有人设计：老年人的室外空间不应该被误解为针对某一特定族群的解决方案，即使为老年人需求定制的环境建设对他们的生活质量至关重要，仍有一定的风险，即让人们觉得自己受到了侮辱歧视。

健康服务的功能被设计为面向所有人群，主要服务半径为周边社区。许多成功的案例都展示了老人与小孩非凡的亲和力。功能的布置绝对不是照本宣科，而是细化到了具体的人群需求：一个中年家庭在上医院看望照顾老人的时候往往是全家出动，缺乏耐心的小孩经常会提前终止看望时间。鉴于此，设置适当的托儿中心，可以增加探望时间，改善老人的心理健康。

Design notes

Design for all: Outdoor Spaces for the elderly should not be misunderstood as a collective solution for a particular group. Even though an environment tailored to the needs of the elderly is critical to their quality of life, there is also a risk that people will feel stigmatized.

The function of the health service is designed to reach all the population, and the main service radius is the surrounding community. Many successful cases show the extraordinary affinity between the elderly and children.

The layout of functions is definitely not descriptive, but refined to the specific needs of the crowd: when a middle-aged family visits and takes care of the elderly in the hospital, the whole family often comes out, and impatient children often terminate the visit time in advance. In view of this, setting up proper child care centers can increase visiting time and improve the mental health of the elderly.

● 前期思考　Early thinking

思考：

如何能把建筑布置得兼具功能性和体验性？

如何构建一种可以产生一系列关系的结构？

建筑内部的使用者如何描述这个建筑以及建筑反过来如何描绘使用者的故事？

场地地处洋湖坑片区，以大河西先导区的起步为建设背景，将建设成中部地区集总部经济区、城市湿地公园、旅游休闲、生态宜居等多种功能为一体的示范城区。洋湖片区总用地 30.5km²，城市建设用地占 97.14%，其中湿地面积占 33.66%。

Think about:
How can the layout of the building be both functional and experiential?
How to build a structure that generates a set of relationships?
How does the interior user describe the building and the building in turn describe the user's story?
The site is located in Yanghu Keng area, with the construction background of the pilot area of Dahexi as the construction background, and will be built into a demonstration city in central region integrating the headquarters economic zone, urban wetland park, tourism and leisure, ecological livable and other functions. The total land area of Yanghu Area is 30.5 square kilometers, 97.14% of which is urban construction land, 33.66% of which is wetland area.

交通流线

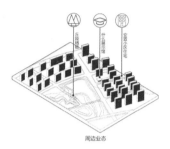

周边业态

● 形体生成　Shape generation

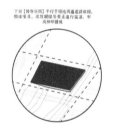    

● 模型照片　Model photos

构建一个有生命力的疗养院的力量源泉，来自常识或自然的力量。所有这些力量，需要被转化为一种形式，而这种形式最终打造的并不是水泥、砖块或者木材，而是生活本身。

The power to build a living sanatorium comes from common sense, or the power of nature. All of these power need to be transformed into a form that ultimately builds not concrete or bricks or wood but life itself.

 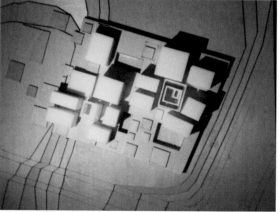

● 立面推敲　Facade deliberation

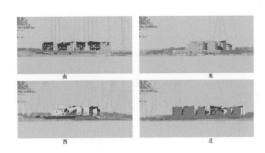

立面推敲：

综合考虑日照，推理单元组合等实际物理因素，对养老病房的组团进行调整、组合。同时辅助以日照软件分析，调整里面开窗与阳台设置。

Facade refinement:
Taking sunshine, combination of reasoning units and other physical factors into consideration, the group of nursing wards is adjusted and combined. At the same time, assist sunshine software analysis, adjust the Windows and balcony Settings inside.

● 剖面设计　Section design

剖面设计：

结合地形高差，通过不同的庭院来对建筑流线进行节奏的控制。结合楼栋之间的围合与走廊的空间组合，营造丰富的聚落空间。

Section design:
Combined with the terrain elevation difference, the rhythm of the building streamline is controlled through different courtyards. Combine the enclosure between buildings and the spatial combination of corridors to create rich settlement space.

模糊立面：

大量的活动和场景在中庭发生，线性的活动，经过二维叠加，活动之间发生交流与对视。

Blurred elevations:
A large number of activities and scenes take place in the atrium. Linear activities, through two-dimensional superposition, conduct exchange and eye contact among activities.

引树入室：

对于需要有更多照顾的介入型调理的老人，在行动上不便让这个群体缺乏足够的倾听外界的机会。所以，在设计中将花、木、湖景引入室内。听雨、赏花、观湖、抚草等体验场景植入其中，刺激老人各方面已经严重衰退的感官。

Tree entry:
For the elderly who need more care for interventional conditioning, the mobility of the group is not enough to listen to the outside world. So in the design of flowers, wood, lake scenery into the interior. Enjoying to the rain, flowers, lake, grass and other experience scenes implanted to stimulate the elderly in all aspects of the serious decline of the senses.

场景演绎·局部空间　Local space

医疗部分的内部空间系统的建立以及外向空间的联系被作为设计的首要出发点。本设计试图消解底层空间的无意义廊道，空间之间的连续性由不同的、有具体功能的房间联系在一起。

The establishment of the internal space system of the medical section and the connection of the external space are taken as the primary starting point of the design. The design attempts to eliminate the meaningless corridors of the ground floor space, and the continuity between spaces is linked by different rooms with specific functions.

场景演绎·鸟瞰图　Aerial view

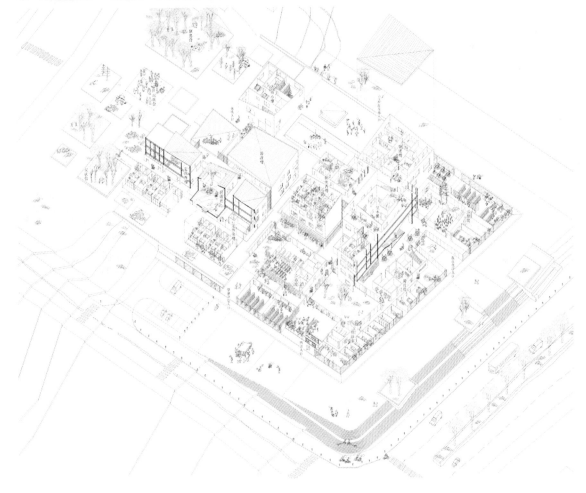

专题四：校园建筑设计
Topic 4: Campus Architecture Design

鸟瞰图

正视图

设计题目： 未来时代的大学空间
指导老师： 彭智谋
学　　生： 张安丽

8

Design topic: University Space in the Future

Instructor: Peng Zhimou

Student: Zhang Anli

● 设计说明

设计提取山脉的曲线线条引导"环"的流动曲线的形态设计，又提取湖湘书院的整体合院式的布局形式，引导建筑南北向的较为规整的布局形式。将两者进行结合，形成了设计概念，引导了初步设计的规划布局雏形，突出了自由与严谨、曲线与方体块的对比与融合。

Design notes

The design extracts the curve lines of mountains to guide the shape design of the flow curve of the "ring", and extracts the layout form of Huxiang Academy, which is the whole courtyard style, and guides the more regular layout form of the building from south to north. The combination of the two forms a design concept, leading to the preliminary design of the planning and layout of the prototype, highlighting the contrast and fusion of freedom and rigor, curve and square block.

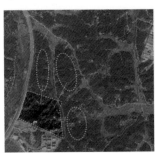
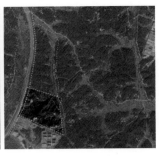
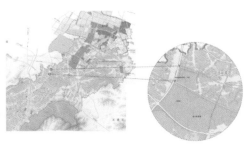

● 地块分析　Plot analysis

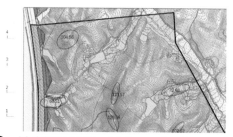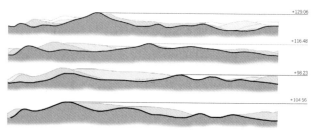

● 城市设计方案　Urban design scheme

总用地面积 41.59 万 m²，含道路代征面积 15220m²，学校建设用地不超过 40 万 m²。基地内部山体陡峭，三面环山，仅有西侧一条市政道路，考虑作为地块主入口，未来拟在北侧扩建一条道路，作为次入口方向。

The total land area is 415900 square meters, including 15220 square meters of road requisition area. The school construction land is no more than 400000 square meters. The base is surrounded by the mountains on three sides. The mountain inside the base is steep. There is only a municipal road on the west side of the site, which is considered as the main entrance of the site. In the future, it is planned to expand a road on the north side as the secondary entrance direction.

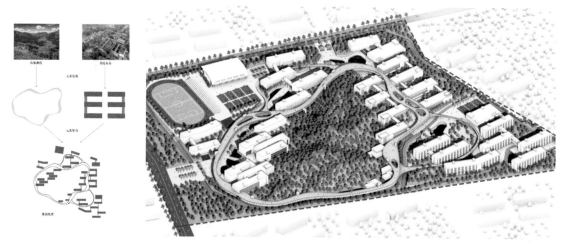

● 结构分析　Structure analysis

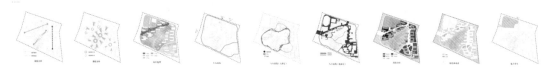

● 剖面图　Profile

剖面1-1

● 场景展示　Scene display

当前许多校园绿化植物种类丰富繁多，但往往未能将环境美化纳入整体规划中，重心都放在楼宇、道路等基础设施建设上，造成校园环境改造的不协调。因此，将基地内现存的大量自然绿化、山体及水体与建筑相结合，形成共生和谐的校园氛围。

At present, many campus are rich in variety of green plants, but often fail to integrate environmental landscaping into the overall planning, but focus on building, road and other infrastructure construction, causing the incongruity of the transformation of campus environment. Therefore, the existing large amount of natural greening, mountain and water in the base are combined with the buildings to form a symbiotic and harmonious campus atmosphere.

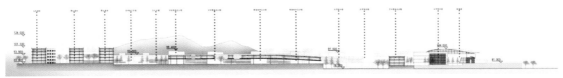

剖面2-2

● 建筑单体 · 平面图　Single building · Plan

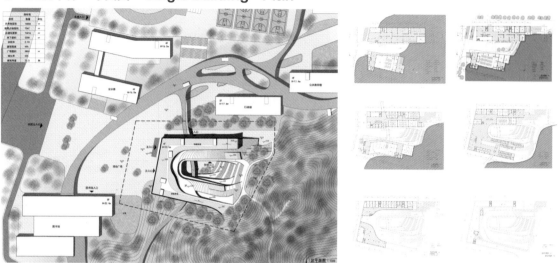

本地块位于湖南省长沙市宁乡市沩东新区，是宁乡的生态新城，也是长沙市的卫星城区，距长沙河西核心梅溪湖片区仅20km，是长沙通往湘中、湘北之要冲，沟通湘西北的咽喉要地。本项目用地性质为A31类（高等院校用地）。南侧同为A31类用地，东、西、北侧为一般林地和基本农田。

The area is located in Weidong New District of Ningxiang City, Changsha City, Hunan Province. It is the ecological new town of Ningxiang city and the satellite urban area of Changsha city. It is only 20 kilometers away from Meixi Lake area, the core of Changsha River west, and it is the key point from Changsha to central and northern Hunan and the key point for communication with northwestern Hunan. The land use nature of the project is A31 (land for colleges and universities). The south side is the same as A31 land, and the east, west and north sides are general forest land and basic farmland.

● 剖透视　Section perspective

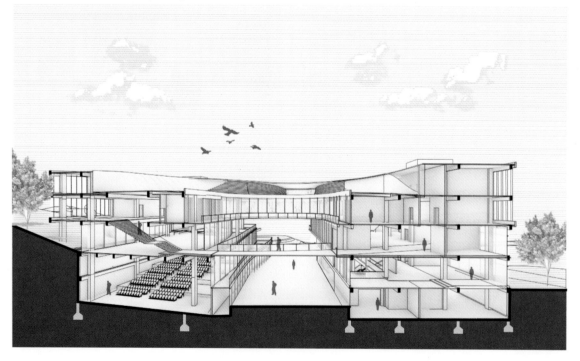

局部效果图

整体鸟瞰图

设计题目：文昌阁小学改扩建设计
指导老师：苗欣
学　　生：任意

9

Design topic: Design of Renovation and Expansion of Wenchangge Primary School

Instructor: Miao Xin

Student: Ren Yi

● **设计说明**

旅游大潮冲击下的历史城镇——凤凰，在商业化的巨大冲击下，历史悠久小学校所面临的困境与策略：一方面，名校的巨大吸引力导致学生人数激增与校园规模的原静态界定方法的冲突；另一方面，外来人口进驻片区也给学校带来难以应对的僵局——地价激增；陪读等流动人口的入住带来的空间需求阻碍校园空间生长；原有校舍在新的教学内容调整上不可持续……这些促成我们重新审视共享时代旅游村镇中的历史名校的空间再组织，从学校内外、周边以至区域上审视当下这一广泛存在的问题，基于从校园人群、陪读人群、服务人群以及旅游人群的冲突中寻求方向，给出合理的校园空间组织。

Design notes

Phoenix, a historical town under the impact of tourism tide, is facing difficulties and strategies under the huge impact of commercialization. On the one hand, the great attraction of famous schools leads to the conflict between the surge in the number of students and the original static definition method of campus scale; On the other hand, the entry of foreign population into the area has also brought about an intractable impasse for schools – a surge in land prices; The space demand brought by the stay of floating population such as accompanying students hinders the growth of campus space; The original school buildings are unsustainable in the adjustment of new teaching contents... These have led us to re-examine the spatial reorganization of historical famous schools in Tourism Villages and towns in the era of sharing, examine the current widespread problems inside and outside the school, around the school and even in the region, and seek direction from the conflict among campus crowd, accompanying crowd, service crowd and tourist crowd, Give a reasonable campus space organization.

凤凰古城风景名胜区核心景区外围

● 调研分析　Investigation analysis

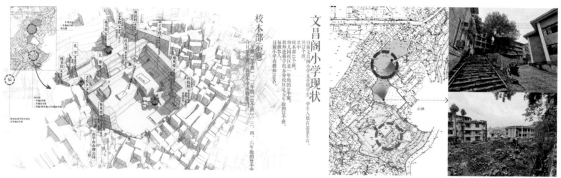

老校区本部是在从前文昌阁的基础上修建起来的，因此园区整体暗示着老的文昌阁（即左上角礼堂、老校区）这条轴线，同时因为曾经的阎王殿（右下角）被用来创办了凤凰县最早的民办中学，后来被拆穿，阎王殿剩余部分被作为校本部的次入口，所以整体校园还暗含着阎王殿这条轴线。

Old campus headquarters is built on the basis of it before, so the overall staple in the old it (that is, the upper left corner hall, the old campus) the axis, at the same time, because once the palace of hell (bottom right) is used to set up the wind burn county to dye the earliest private middle school, was caught later, palace of hell rest as school cadre's entrance, So integral campus still implies yan Wangdian this axis.

● 场地认识模型　Site cognition model

● 方案进程　Programme process

 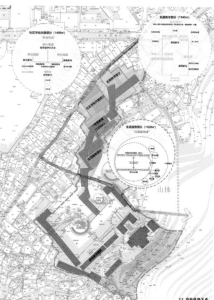

一草阶段主要考虑的是场地意象的延续，对于场地的利用相对局限。二草阶段将一草时的模型打散，在察觉出一草有一种阻隔意味的态势时，重新思考场地中建筑希望呈现的态势。

The first draft stage mainly considers the continuation of the site image, and the utilization of the site is relatively limited.

In the second grass stage, the model of one grass is broken up. When the situation of one grass has a barrier meaning, rethink the situation that the buildings in the site want to present.

总平面图　General layout

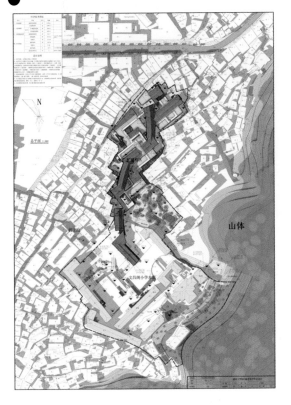

主要通过景观设计及部分加建，梳理原有校园轴线，依据调研对现有建筑及校园场地进行评价，决定构筑物的增设，协调旧校园的空间矛盾，调整校园空间，使扩建部分应对当下校园空间不足的问题及对未来拓展校园空间的需求。

通过对山地地形处理和种植平台的设置，合理布置改善性消防车道，同时使建筑肌理与古城社区肌理相协调。基于对当下及未来的教学空间的拓展及校园服务结构完备的需求，设综合服务楼建筑面积约 6000 ㎡（架空层部另算）。

Mainly through the landscape design and expansion, combing the existing campus axis, based on the research to evaluate existing buildings and campus space, structures, decide the added coordinate the contradiction of the old campus space, adjust the campus space, make the expansion part of the deal with the problem of shortage of the current campus space and the demand of the campus space for future expansion.

Through the treatment of mountain terrain and the setting of planting platform, the reasonable arrangement of the fire lane, and the fabric of the building is coordinated with that of the ancient city community. Based on the current and future expansion of teaching space and the demand of complete campus service structure, the comprehensive service building covers an area of about 6000 square meters (the overhead floor is calculated separately).

关系分析　Relationship analysis

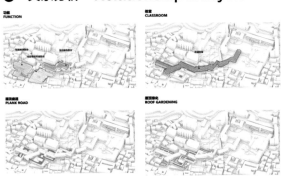

基座式的平屋面屋顶设置轻质植草层，考虑使得整个建筑的体量及肌理与周围相契合，同时也为顶层教室带来自然的漫反射采光。充分利用基座式的平屋面屋顶设置屋顶廊道，连接于教室或者教室外走廊间，为每个班级学生就近提供课间活动场所。散落在各个部分的种植园为学生提供认识草药、探索自然的校园活动，而在读书交流部的种植空间为学生从认识到实践交替探索知识的场所。老校区中置入了部分小型种植园及观察棚构筑物。

The plinth flat roof is equipped with light grass layer, which makes the volume and texture of the whole building fit with the surrounding area, and also brings natural diffuse lighting to the classroom on the top floor. Make full use of the flat roof of the pedestal to set up a roof corridor, which is connected to the classroom or the corridor outside the classroom, providing the nearest place for students of each class to have recess activities. The plantations scattered throughout the sections provide campus activities for students to learn about herbs and explore nature, while the planting space in the reading exchange section provides a place for students to explore knowledge alternately from knowledge to practice. Some small plantations and observation sheds were placed in the old campus.

流线分析　Streamline analysis

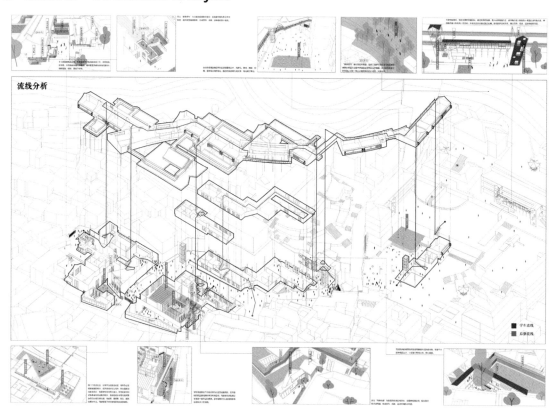

功能布局　Functional layout

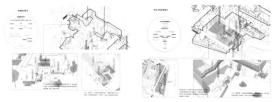

手工教室搭配阅读角，配置有关蜡染等少数民族传统工艺、相关植物的书籍，外侧有蜡染植物种植园，植物配置为蜡染染料的原材料植物蓝靛、椿树、黄栀子树等。在以"亲手劳作"为主题的拓展教学部分，设置读书角，与手工作坊教室、自然教室连接起来，形成劳作、阅读、运动相间的小组团。

The manual classroom is equipped with reading corner, batik and other minority traditional crafts and related plants books, and outside are the batik plant plantation, which includes batik dye raw materials plants indigo, toon tree, yellow gardenia tree and so on. In the extended teaching part with the theme of "working with one's own hands", reading corners are set up to connect with manual workshop classrooms and nature classrooms to form groups with work, reading and sports.

台地形成了湘西地区特色运动如苗族武术、高脚马、舞龙、舞狮、陀螺、苗拳等活动的场地，围合的廊道成为观众席，场地成了舞台。在老校区部分，结合主要的交通空间，通过向外的拓展，置入以种植园为主、读书角和种植园为辅的空间，在老校区部分通过整合处理，形成教学空间为主，辅以劳作、阅读、运动相间的组团。

The bench terrace forms the venue for the special sports in western Hunan, such as Miao wushu, steed horse, dragon dance, lion dance, gyro, miao boxing, etc. The enclosed corridor becomes the auditorium and the venue becomes the stage. In the old campus, combined with the main traffic space, through outward expansion, we put into the reading corner as the main part, and the space with the plantation as the auxiliary. In the old campus, through integration, there formed the teaching space as the main, which is supplemented by work, reading, sports group.

● 剖透视　Section view

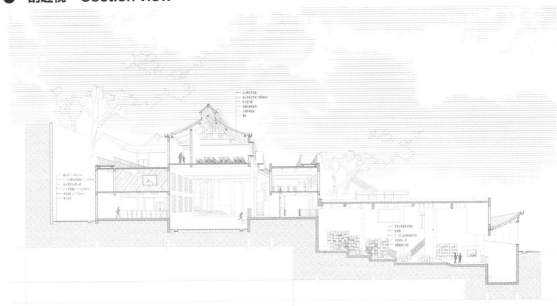

● 效果图　Rendering

通过改扩建之后，校园整体态势以及学校与社区的共建关系，以悠久的学校历史和稳定的社区记忆带来的力量在有限的范围内抵御古城景区商业文明扁平化的扩张。

Through the overall situation of the campus after the renovation and expansion and the co-construction relationship between the school and the community, the power brought by the long school history and stable community memory can resist the expansion of commercial civilization flattening in the ancient city scenic spot within a limited scope.

专题五：公共文化建筑设计
Topic 5: Public Culture Architecture Design

沿街城市景观

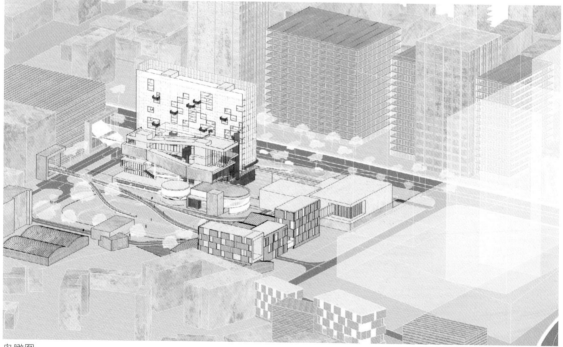

鸟瞰图

設计题目：水草之家——新工人文化时代文化宫重构
指导老师： 向昊
学　　生： 于童

10

Design topic: Home of Aquatic Plants —— Reconstruction of Cultural Palace in the New Workers' Cultural Era

Instructor: Xiang Hao

Student: Yu Tong

● 设计说明

本项目为原工人文化宫翻新改造，场地面积大，路网密度低，功能类型单一。地处城市商业地域，对场地的利用强度和功能程序提出了要求，应努力接纳周边功能，促进地块多元化利用。故将场地分为多个地块，设置不同功能，同时保留原有建筑，丰富场地风貌。

文化宫是特殊历史时期的产物，与工人社区联系紧密，它与当下社会的文化需求有所区别。如何维持公益性，维持文化的自更新自生产，提供融入地域生活圈的新型文化建筑，是本次设计的探索方向。

Design notes

This project is the renovation of the former Workers' Cultural Palace, with large site area, low density of road network and single function type. Located in the commercial area of the city, the utilization intensity and functional procedures of the site are required. Efforts should be made to accept the surrounding functions and promote the diversified utilization of the site. Therefore, the site is divided into multiple blocks to set up different functions while retaining the original buildings to enrich the site style.

The Cultural Palace is the product of a special historical period, closely connected with the workers' community, and it is different from the cultural needs of the current society. The exploration direction of this design is how to maintain the public welfare, the self-renewal and self-production of culture, and provide new cultural buildings integrated into the regional life circle.

1970年代老长沙地图
片区为工人宿舍区（深色部分）

主要工人住区
片区为工人宿舍区
但仍未搬迁的老工人不过半数
大量务工者、学生家长涌入

适老活动场所
以社区公园和棋牌室为主
文化生活相对单一
活动场所相对缺乏

■工人文化分类

老工人文化	新工人文化	打工者文化
改制前国企为主的 工人文化	改制后 工人文化	大量务工者 个体经营者文化

■老工人文化

老工人文化者曾是文化宫的利用者，但随着年岁的增长超龄脱团现象明显
文化宫曾留下其印记与记忆
提示可能引入相应的适合老工人之活动或福祉服务
使老工人重活文化舞台之席位

● 调研分析　Investigation analysis

兴起于 20 世纪 80 年代，以百货贸易为基础，加之初步繁荣的个体商贩，在工人宿舍集中的片区内初具影响力。

Rose in the 1980s, based on the department store trade, coupled with the initial prosperity of individual vendors, it has begun to have influence in the area where works' dormitories are concentrated.

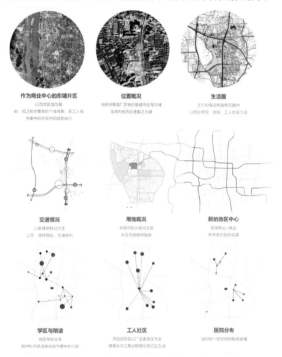

● 群体策略　Group strategy

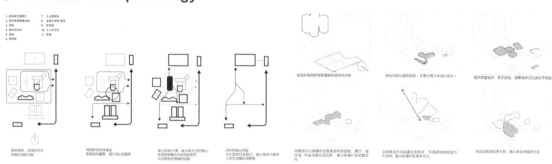

根据前期调研、区域城市设计和概念设计指导，拟建设一个微型公寓与文化建筑的混合体。场地内同时规划有其他建筑：南侧城市综合体的建设将为场地带来稳定的人流，结合南侧的巴比伦溜冰城以及东侧大学，设置歌舞娱乐场所、夜市、文创设施等，将其转变为连接地铁站的商业后街的一部分。同时将场地内水塘部分作为洼地的特色保留，修整自然景观，并引导商业人流穿越连接至北部商贸城，联通门前的商业地下街，形成体验丰富的循环动线。

According to the preliminary investigation, regional urban design and conceptual design guidance, it is proposed to build a complex of micro apartments and cultural buildings. At the same time, other buildings are planned in the site: the construction of urban complex on the south side will bring stable flow of people to the site. In combination with the Babylon Skating City on the south side and the university on the east side, singing and dancing entertainment venues, night markets, cultural and creative facilities will be set up, transforming it into a part of the commercial back street linking the subway station. At the same time, the pond in the site is retained as the feature of the depression, and the natural landscape is repaired, and the commercial flow of people is guided through the commercial underground street connected to the northern trade city, forming a circular line with rich experience.

● 群体与城市关系　Relationship between groups and cities

可照时间计算

视廊与体量

● 形体生成　Shape generation

新工人阶级的社区，引入一个空间的竞争者，以小时为单位的领域进退与相互转化，社区功能每天重置缓冲区，防止文化宫的固化。

The new working class community introduced in a spatial competitor, advancing and retreating and transforming each other on an hourly basis, so as to prevent the solidification of the cultural palace through the buffer zone.

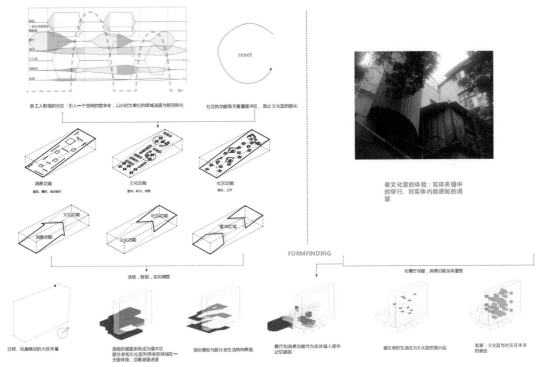

● 运行机制　Operating mechanism

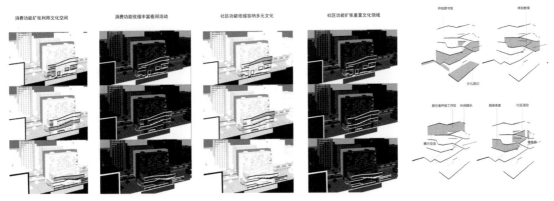

● 场地与社区　Site and community

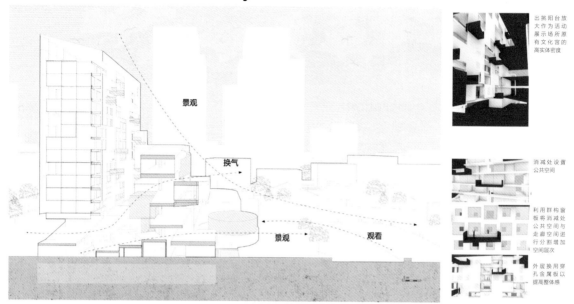

出挑阳台放大作为活动展示场所原有文化宫的高实体密度

消减处设置公共空间

利用群构窗板将消减处公共空间与走廊空间进行分割增加空间层次

外层换用穿孔金属板以提高整体感

● 坡道系统与功能　Ramp System and function of the ramp

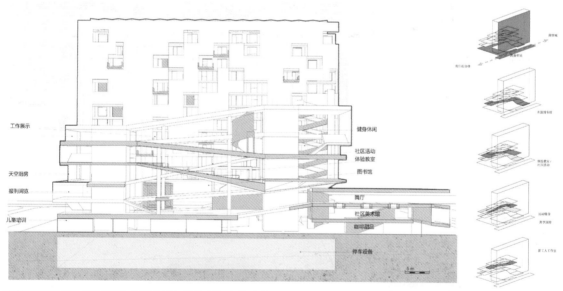

连续的坡道与阶梯系统形成缓冲区，居住者和文化宫使用者的领域在一天中转换，沿着坡道进退，建筑因而是实时、瞬息可变的。这种领域的转换取进退之意。

在工作时间内这个系统作为公共的文化建筑被使用，而当晚上到来，这些区域可作为新工人工作和学习的拓展场所。这个过程同时完成对文化宫的重置，避免文化宫内消费功能的无限扩张最终造成程序和使用人群的固化。

A continuous system of ramps and steps serves as a buffer zone, and the areas of residents and users of the cultural palace change throughout the day, advancing and retreating along the ramps, so that the building is real-time and instantly changing. This shift means advancing and retreating.

The system is used as a communal cultural building during working hours, but in the evening, these areas can be used as a place for new workers to work and study. This process also completes the resetting of the Palace of Culture and avoids the infinite expansion of consumption functions in the Palace of Culture, which eventually leads to the solidification of programs and users

● 窗景与对望　Window view and opposite view

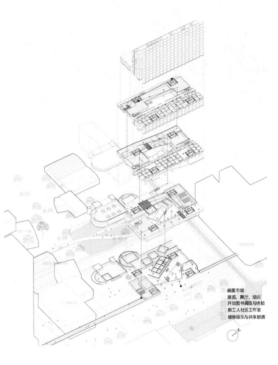

拼盖市场
展览、舞厅、培训
开放图书阅览与体验
新工人社区工作室
健身娱乐与共享厨房

● 效果图　Rendering

半透明的表皮分隔公寓系统和城市，其开启状态提供了与城市界面间距离控制的手段，使阳台成为生活的场所和活动的微型舞台。面向东塘的一面保留了场所的特质，由南侧渗透进入的商业人流在这里体验宁静，文化部分—公寓—自然间提供了多种层次和对望关系。

The translucent skin separates the apartment system from the city, and its open state provides a means of controlling the distance from the urban interface, making the balcony a place of living and activities. The side facing the East Pond retains the character of the place, where the commercial flow from the south side penetrates to experience tranquility, while the cultural part – apartment – nature provides multiple levels and views.

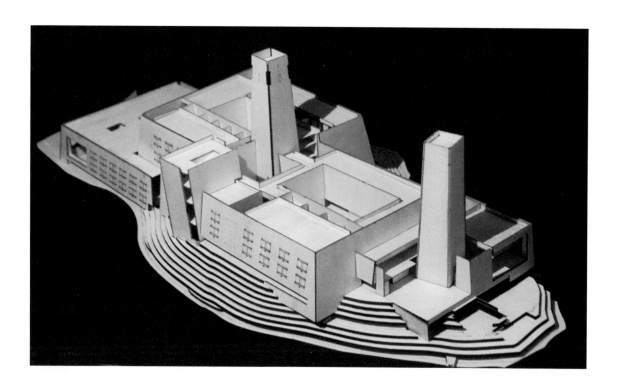

模型展示 1

模型展示 2

62

11

设计题目： 瞭望与庇护——松岗县官寨禅修中心设计
指导老师： 蒋甦琦
学　　生： 李维衡

Design topic: Lookout and Shelter —— Design of Guanzhai Meditation Center in Songgang County

Instructor: Jiang Suqi

Student: Li Weiheng

● **设计说明**

基地处于几条交通要道之间，且高出各条路近百米的高度，本身具有强烈的观望属性；另外，基地中本来的官寨就是一所"具有观望台的庇护所"，而且碉楼作为观望的代表遗留了下来。因此，选取观望与庇护作为概念，用现代语言复述碉楼的故事。

由分析得到，由于基地基础设施问题，本地不适合作为一般过路旅客的消费旅社，而相对适合作为长期休假、逃离城市的度假之地。又由于藏区宗教氛围浓厚，且在世界范围内已有的禅修中心多建在景观开阔且有水系的基地，因此本处功能设定为禅修中心。

Design notes

The base is located between several major traffic roads and is nearly 100 meters higher than each road, which has strong observation property. In addition, the original Guanzhai in the base is a "shelter with observation tower", and Diaolou is left as a representative of observation. Therefore, the concept of lookout and shelter is selected to retell the story of Diaolou in modern language.

According to the analysis, due to the infrastructure problems of the base, there is not suitable as a consumer hotel for general passing tourists, but relatively suitable as a vacation place for long-term vacation and escape from the city. In addition, due to the strong religious atmosphere in Tibetan areas, and the existing meditation centers around the world are mostly built in the base with open landscape and water system, so the function of this office is set as a meditation center.

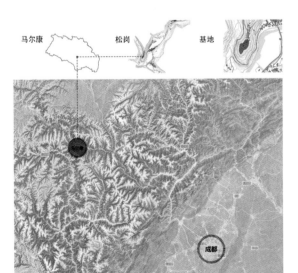

马尔康　　　松岗　　　基地

成都

● 场地分析　Site analysis

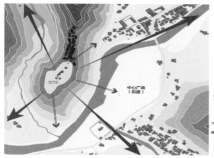

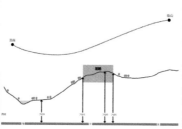

建筑基地处于峡谷中一个山丘的山顶，一侧背山，三侧临山谷，有很强的观望感与独立感。根据"观望与庇护"原理，这里适合修养或者禅修。

The construction base is located at the top of a hill in the canyon, flanking the mountain and facing the valley on three sides. It has a strong sense of outlook and independence. According to the principle of "lookout and shelter", it is suitable for self-cultivation or meditation.

● 元素提取　Element extraction

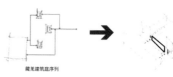

汉式建筑庭院通常被建筑群的正立面主动围合出来，这样建筑庭院与建筑就是从属关系，而不能作为一个独立的元素纳入公共生活。而在藏羌建筑聚落中，经常发现一些类似于被建筑被动围合出来的庭院。由于没有任何建筑与之产生序列关系，因而处于主要的通路上，从而成为独立于房间的存在而非房间的延伸。

The Han style architectural courtyard is usually actively enclosed by the front facade of the architectural complex, so that the architectural courtyard is subordinate to the building and cannot be incorporated into the public life as an independent element. In Tibetan and Qiang architectural settlements, some courtyards are often found to be passively enclosed by buildings. Since no building has a sequential relationship with it, it is in the main path and becomes an existence independent of the room rather than an extension of the room.

● 构成逻辑　Constituent logic

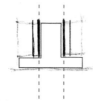

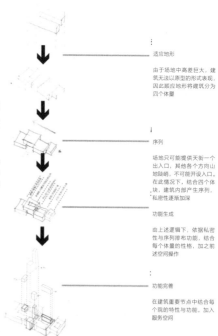

在现有模式下，套型围合出的公共空间呈现出三种主要状态：中间公共院子里的全公共状态；剪力墙和院子围墙之间的半公共状态和剪力墙之后可以作为较私密公共功能也可以作为公共功能的服务空间。

Under the existing mode, the public space enclosed by the condominium presents three main states: the whole public state in the middle public yard; The semi public state between the shear wall and the courtyard wall and the rear of the shear wall can be used as a more private public function or as a service space for public functions.

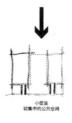

● 套型讨论 Suite discussion

套型原型

通过上下联通的卫生间（仅容一个马桶）分割整个空间，空间与卫生间体量的距离决定空间的性质与功能。同时在"院"区通过不同天窗的关系，在给院子带来下层公共性的同时，形成了特殊的空间效果。

Suit prototype
The whole space is divided by the toilet connected up and down (only one toilet). The distance between the space and the toilet volume determines the nature and function of the space. At the same time, through the relationship between different skylights in the "courtyard" area, the special space effect is formed while bringing the lower level publicity to the courtyard.

【 套型讨论1：禅修&VIP 】   【 套型讨论2：安养空间 】

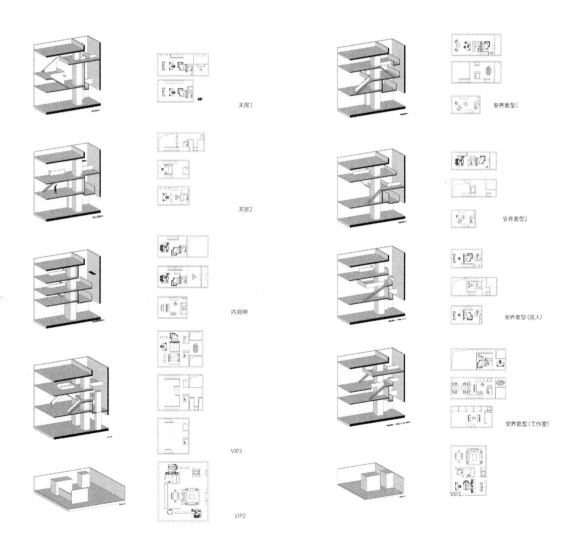

关房1

关房2

内观禅

VIP3

VIP2

安养套型1

安养套型2

安养套型（双人）

安养套型（工作室）

VIP1

● 总平面图 · 功能流线　General layout · functional streamline

项目基地位于马尔康市松岗县，对大多数旅行者而言属于长途旅行范畴，到此的游客大都拥有较长且弹性的旅行时间，且基地正好处于一天车程的中转站，将其设计为旅馆具有优势。

The project base is located in Songgang County, Malkang City, which belongs to the category of long-distance travel for most travelers. Most of the tourists here have a long and flexible travel time, and the base is just located in the transfer station of a day's drive, so the base has an advantage in developing the hotel industry.

● 节点效果　Node effect

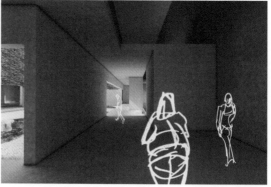

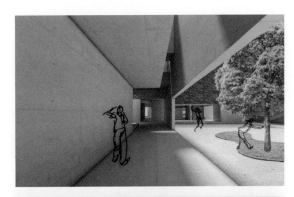

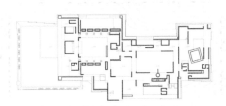

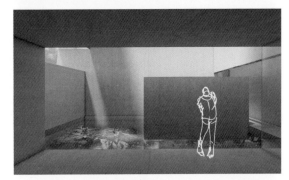

建筑入口透视

整体效果图

12

设计题目：官寨遗址文化体验博物馆设计
指导老师：蒋甦琦 邓广
学　　生：赵亮

Design topic: Design of Cultural Experience Museum of Guanzhai Site

Instructors: Jiang Suqi, Deng Guang

Student: Zhao Liang

● 设计说明

该项目是一个主要用于参观体验的建筑，同时配备村民服务设施及相关配套用房的开发项目。设计中，首先充分考虑了场地现有人流状况，分析了地形高差，合理布置交通出入口；其次，设计试图改变传统博物馆脱离日常生活、忽略公共交往活动的现状。尝试在现有空间平面的限制下，实现顶层置入公共空间的实验，并通过设置单独的路径分别串联游客展厅以及村民活动空间，激发高层建筑内部空间的活力；通过环形上升的参观流线与直接的村民流线的叠合，激发博物馆内部垂直功能的联系与交流，实现情景交融，增强体验。

Design notes

The project is a development project which is mainly used for visiting and experiencing buildings, and also equipped with villager service facilities and related supporting houses. In the design, first of all, the existing flow of people in the site was fully considered, the terrain height difference was analyzed, and the traffic entrance and exit were reasonably arranged. Secondly, the design attempts to change the status quo that traditional museums are divorced from daily life and ignore public communication activities. Under the restriction of the existing space plane, it tried to realize the experiment of placing public space into the top floor, and set paths to connect the tourist exhibition hall and the villagers' activity space respectively, to stimulate the vitality of the internal space of the high-rise building. Through the superposition of the circular ascending circulation line and the direct villager circulation line, the connection and communication of the vertical function inside the museum is stimulated to realize the integration of scene and enhance the experience.

● 前期分析　Pre-phase analysis

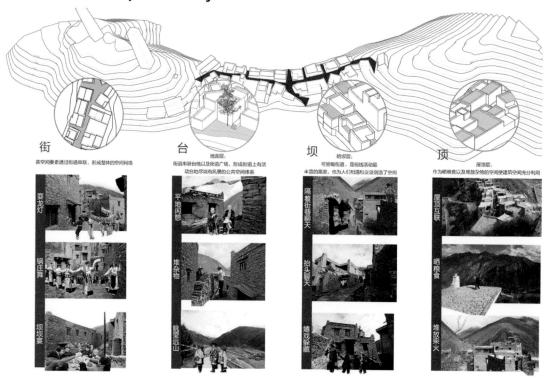

街

各空间要素通过街道串联，形成整体的空间网络

歪龙灯

锅庄舞

坝坝宴

台

地面层；
街道串联台地以及街道广场，形成街道上有活动台地尽端有风景的公共空间体系

平地闲憩

堆杂物

眺望远山

坝

晒坝层；
可俯瞰街道，是视线活动最丰富的高度，也为人们相遇和交谈创造了空间

隔着街相聊天

抬头聊天

嬉戏躲藏

顶

屋顶层；
作为晒粮食以及堆放杂物的空间使建筑空间充分利用

屋顶互联

晒粮食

堆放柴火

调研结论：

不论在过去还是现在，在日常还是节庆，天街都是当地人生活的纽带或载体；

天街生活的激活是有必要的；

天街上"紧凑""高低"的空间关系为生活行为塑造了充分条件；

天街的空间特色应该在新的建筑基地内以新的方式重生。

Research conclusion：

In the past and now, daily and festive activities，Tianjie is the link or carrier of local life；

It is necessary to activate life in Tianjie；

The "compact" and "high and low" spatial relations on Tianjie create sufficient conditions for living behaviors；

The spatial characteristics of Tianjie should be reborn in a new way within the new building site.

● 策略思考　Strategic thinking

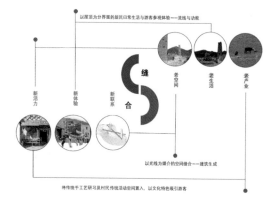

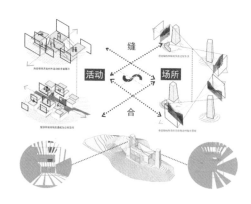

● 形体生成　Shape generation

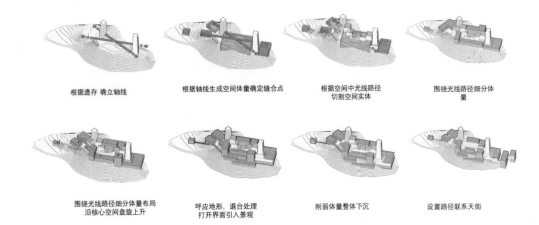

根据遗存 确立轴线　　根据轴线生成空间体量确定缝合点　　根据空间中光线路径　　围绕光线路径细分体
　　　　　　　　　　　　　　　　　　　　　　　　切割空间实体　　　　　　量

围绕光线路径细分体量布局　　呼应地形，退台处理　　削弱体量整体下沉　　设置路径联系天街
沿核心空间盘旋上升　　　　打开界面引入景观

● 总平面图　General layout

建筑周边主要人流来自天街方向，故将博物馆主入口对接天街设计，并通过垂直交通迅速提升至参观标高平面。

The main flow of people around the building comes from the direction of Tianjie, so the main entrance of the museum is connected with the design of Tianjie, and quickly raised to the visiting elevation plane through vertical traffic.

● 功能流线　Functional streamline

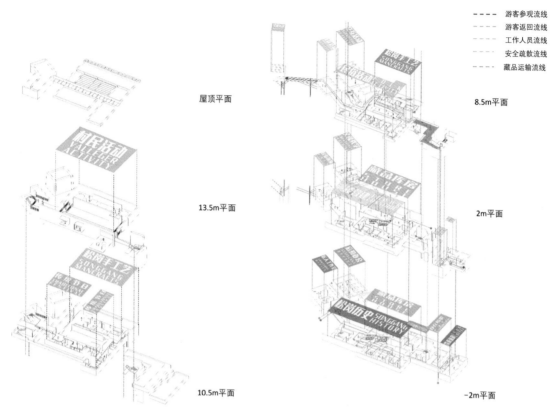

屋顶平面

8.5m平面

13.5m平面

2m平面

10.5m平面

-2m平面

通过联通场地北侧与天街的景观集散广场来组织外部人流关系；主入口设置于北侧电梯口，并通过室外廊桥与室内坡道介入室内环形流线。场地南侧考虑疏散，设置步行道延伸至山下，与北部天街实现人流南北分散。

Organize the external flow of people through connecting the landscape distribution square of the north side of the site and the Sky Street; The main entrance is set at the elevator entrance on the north side, and through the outdoor corridor bridge and the indoor ramp to join the indoor circular flow. The south side of the site is considered to set up a walking path to extend to the foot of the mountain, and realize the north-south dispersal of people in the north of Tianjie.

● 剖面图　Profile

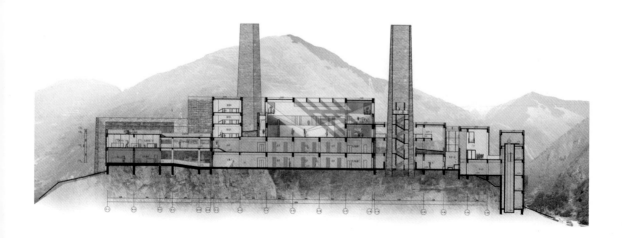

● 效果图　Design sketch

设计构思图

整体鸟瞰图

13

设计题目： 界限——铜官窑展示聚落设计
指导老师： 蒋甦琦 张蔚 向昊 袁朝晖 邓广
学　　生： 王元钊

Design topic: Boundary —— Tongguan Kiln Display Settlement Design

Instructors: Jiang Suqi, Zhang Wei, Xiang Hao, Yuan Zhaohui, Deng Guang

Student: Wang Yuanzhao

● **设计说明**

我们希望在游人见到手艺人之前，手艺人可以先观察到游人的存在，即在受到游人的窥视之前，被窥视者（手艺人）可以提前知道游人的到来。因此，手艺人能得到所需的安全感，甚至可以拒绝游人的观看。如此一来，手艺人不再因为一双不知什么时候出现的眼睛而感到不安，手艺人与游人都似乎处于一个相对平等的位置，平等地交流、平等地接触。我们也衷心希望，能通过设计，用空间或者空间序列，使得这样的交流成为可能。

Design notes

We hope that before visitors see artists, artists can take a step ahead and observe the existence of visitors. That is, before being peeped by visitors, peep (artisans) can know the arrival of visitors in advance; Therefore, craftsmen can get the needed sense of security, and even refuse to be watched by visitors; In this way, the craftsman is no longer troubled by the appearance of a pair of eyes. Visitors are no longer see the impatient eyes of the artisans. Under such conditions, craftsmen and visitors seem to be in a relatively equal position, equal communication, equal contact; We also sincerely hope that through design, space, or spatial sequence, we can make this kind of communication possible.

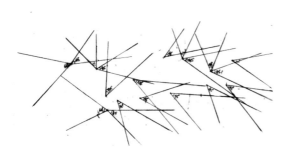

窥视一般定义于从一个小洞窥探

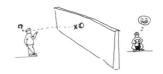

但是，小孔的定义在空间上不存在

倘若单纯地把窥视的定义转换为空间上

平面

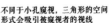

不同于小孔窥视，三角形的空间形式会吸引被窥视者的视线

调研分析 · 多种界线并存的矛盾体
Investigation analysis · a paradox with many boundaries coexisting

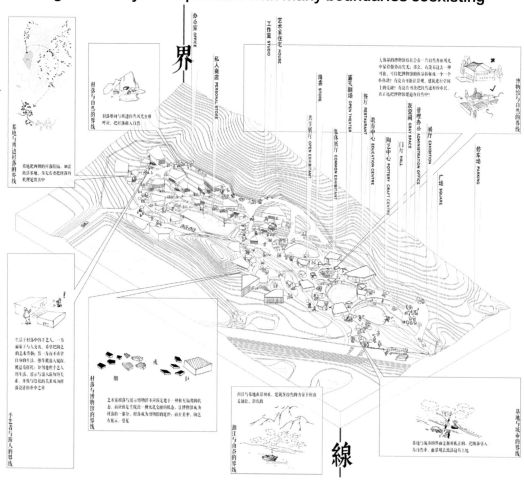

设计手法　Design technique

大体型的博物馆在如此复杂多变的地势中往往会显得冲突而强硬，正如我们把一个人工打造而成的体块生硬地扔在场地上。那么，作为一个博物馆设计，处于秀美的风景中，有没有可能把整体的博物馆空间彻底打碎？有没有可能在这里让建筑空间与景观空间进行一次极致的交融？有没有可能室内与室外的空间实验般地模糊，在游走中不断地转换呢？

Large museums in such a complex and varied terrain tend to be confrontational and forceful, just as we throw a man-made block on the field. So, as a museum design, in a beautiful landscape, is it possible to completely break the whole museum space? Is it possible to combine architectural space with landscape space here? Is it possible that the space between indoor and outdoor is experimental and fuzzy, constantly changing in the walk?

● 平面生成　Plane generation

室内空间与室外空间以一墙为隔，以介乎于室内空间与室外空间的灰空间为媒介，或以穿透室内外空间的不同程度的灰空间为桥梁，连接各个博物馆空间，使得游于其中的游人迷失在室内外空间，让景观融入博物馆中。为了让人在游历的过程中不断地体会建筑、景观、灰空间，以此引起人对于室内、室外空间的界定的模糊感；这三种空间则需要均质、间隔地分布。

由于空间序列和功能体块散落在基地上，均质的空间在村落的生长线、功能面积、视线关系的作用下慢慢形变，生成一个自然的、融合的博物馆平面图。

With a wall of isolation, indoor and outdoor space make the grey space between indoor space and outdoor space as a medium, or the grey space with different degrees of penetration of indoor space or outer space as a bridge to link museum space, so visitors lost in the indoor and outdoor space and let the landscape melt into the museum. In order to make people constantly experience architecture, landscape and grey space in the process of travel, so as to cause people to the definition of indoor and outdoor space fuzzy sense; These three Spaces need to be evenly spaced.Flat functional blocks arranged according to spatial order are scattered on the base, and the homogeneous space is slowly deformed under the action of the village's growth line, functional area and line of sight, generating a natural and integrated museum plan.

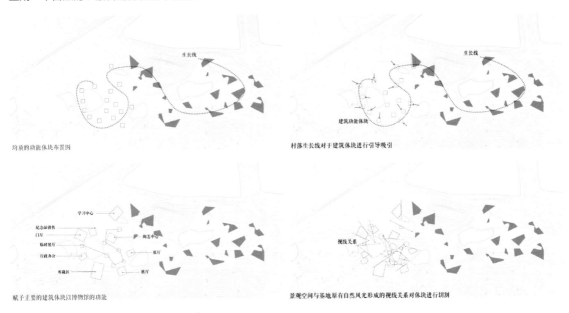

均质的功能体块布置图

村落生长线对于建筑体块进行引导吸引

赋予主要的建筑体块以博物馆的功能

景观空间与基地原有自然风光形成的视线关系对体块进行切割

● 立面设计　Facade design

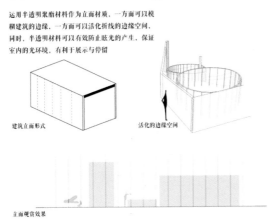

运用半透明聚酯材料作为立面材质，一方面可以模糊建筑的边缘，一方面可以活化拆线的边缘空间，同时，半透明材料可以有效防止眩光的产生，保证室内的光环境，有利于展示与停留。

建筑立面形式

活化的边缘空间

立面硬装效果

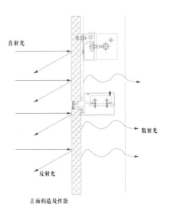

立面构造及性能

● 功能分区　Functional partition

博物馆功能分区

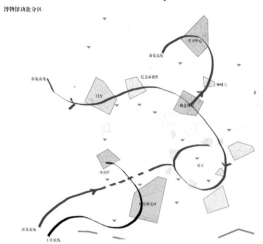

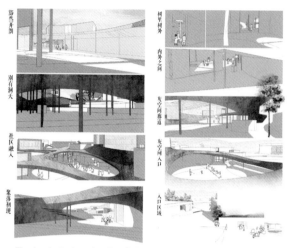

从铜官窑陶器的基本样式出发，简化抽象成一道道曲线、弧线，并以此为基本元素进行墙体设计；如此，人在游历博物馆的过程中，看到不同形式的弧墙、不同形式的陶墙，会产生如同在遗址中寻觅、发现、挖掘铜官窑的惊喜感。

The basic design of pottery is based on the simplification of the wall and the curve of pottery; In this way, when people see different forms of arc walls and ceramic walls in the process of visiting the museum, they will have a pleasant surprise like looking for, discovering and excavating copper official kilns in the ruins.

● 效果图　Rendering

● 模型展示　Model display

基地三面临山，一面临江，一片秀丽风景。然而在基地的山脚，已修建了一座博物馆，强势地占领了一片山林。作为一个联系江与水、游人与艺术村的博物馆，有没有可能把本应巨大体量的博物馆打碎、拆解，与艺术村连接在一起，把室内空间与景观融为一体，把风景、场地还原给市民？

With beautiful scenery, the base faces mountains on three sides and rivers on one side. At the foot of the base, however, museum was built, here dominating a wooded area. As a museum that connects river and water, visitors and art village, is it possible to break up the museum, and link it with the art village, so as to integrate the interior space with the landscape and restore the scenery and site to the citizens?

场地入口透视 1

场地入口透视 2

14

设计题目：长沙铜官窑博物馆
指导老师：蒋甦琦 张蔚 向昊 袁朝晖 邓广
学　　生：张珏

Design topic: Changsha Tongguan Kiln Museum

Instructors: Jiang Suqi, Zhang Wei, Xiang Hao, Yuan Zhaohui, Deng Guang

Student: Zhang Jue

● **设计说明**

Design notes

项目周边自然条件优越，山脉环抱，面朝湘江，基地本身有高大的乔木覆盖，项目将自然融入设计，为建筑创造良好的自然环境，为参观创造良好的景观品质。此外，基地是铜官窑遗址公园的重要节点，如何将新建博物馆合理安插在遗址公园的整条浏览流线内，亦是本案的一个重点。充分利用场地地形，力求组织布局合理，功能分区明确。

The surrounding natural conditions of the project are superior, in that it is surrounded by mountains, facing the Xiangjiang River, and the site itself is covered by tall trees. The project should integrate nature into the design, create a good natural environment for the building, and a good landscape quality for visitors. In addition, the base is an important node of the Tongguan Kiln Site Park, and how to reasonably place the new museum in the whole browsing line of the site park is also a key point of this case. Make full use of the terrain of the site, and strive for reasonable organizational layout and clear functional zoning.

● 调研分析　Investigation analysis

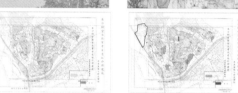

作坊制作　　　　　　　龙窑烧造　　　　　　市场交易　　　　　　码头运输

基地位于铜官窑（长沙窑）遗址公园内，是长沙古窑制造瓷器的原始生产地。古窑造瓷，是因地造窑、就地取材、与自然协作的智慧。长沙窑所在地燃料充足，瓷土资源丰富，水路交通方便。作为长沙窑遗址公园中的博物馆，旨在全面地体现长沙窑遗址所在、产品特点，浓缩长沙窑的故事。这里可以是游览遗址公园的开始，也可以是终点。

Located in the Tongguan Kiln (Changsha kiln) site Park, the base is the original production place of Changsha ancient kiln porcelain. Ancient kiln making porcelain is the wisdom of making kiln according to the place, using materials locally and cooperating with nature. The location of Changsha kiln is abundant in fuel, rich in porcelain clay resources and convenient in maritime transportation. As a museum in changsha Kiln Site Park, it aims to fully reflect the location of changsha kiln site, product characteristics, and condense the story of Changsha kiln. It can be the beginning or the end of the tour of the site park.

● 方案生成　Scheme generation

并置

自然材料—景观路径

陶瓷器物—中心展示空间

人工制作—环绕展示

揉 / 塑

单元

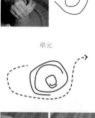

自然元素：
土："挖泥洞"景观 /
夯土平台 / 泥
火：柴 + 砂石
水：水塘 + 碎瓷
草

制作过程：
智泥·画彩·烧制（龙窑）·
聪瑙

陶瓷器物：
第一组：胎——质朴的开始
第二组：釉——色彩系列
第三组：釉——白瓷系列
第四组：釉——字画系列
第五组：器物——造形系列

82

● 功能分区与流线　Functional zoning and streamline

长沙窑是历史上制作陶瓷器
的地方。制作陶瓷器的过程，
是取材于自然制作人造物的
过程，可以把它解读为自然－
器物－人行为的对话关系。
因此，我们在建筑中将自然
因素－陶瓷器物－制作过程
并置。如同制陶中雕塑动作，
将并置的 3 种空间卷曲塑形
成为主体空间单元。

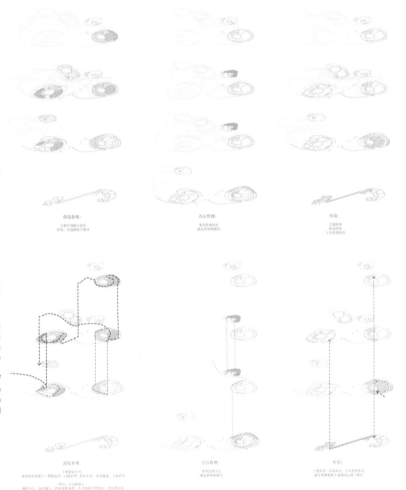

The base Changsha kiln was once
a place where people made pottery
and porcelain. This is the process
of drawing materials from nature,
which can be interpreted as the
dialogue relationship between
nature, utensils and human
behavior. This juxtaposes the
natural elements of the building –
the ceramic objects – the process of
making them. Just like the sculpture
movement in pottery making, the
three spaces are crimped and
shaped into the main space unit.

● 材料选取　Material selection

在建筑材料中，夯土和混凝土是具有模板塑形逻辑、具
有浇筑特点的材料，和制陶时通过模具塑造陶瓷形状的
方法如出一辙。建筑外院落外墙使用夯土墙体。展览主
体使用混凝土筒体与墙承重结构，环绕形成螺旋形展览
空间。

Among the building materials, rammed earth and concrete are the
materials with formwork plastic logic and pouring characteristics. It's
the same as you do when you mold pottery. Rammed earth wall is
used for exterior wall of building yard. The main body of the exhibition
uses a concrete cylinder and wall bearing structure, forming a spiral
exhibition space around.

● 模型照片　Model photos

1：300 模型

主要材料 :3mm 激光切割奥松板、2mm 白色 pvc、3mm 软木卷、透明纸

1:300 model
Main materials :3mm laser cut Auzon board, 2mm white PVC, 3mm cork roll, transparent paper

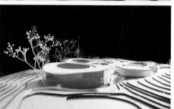

1：50 主展馆 2 中心筒体模型

主要材料：3mm 激光切割奥松板、3mm 软木卷、5mm 软木卷、白色草粉

1：50 main exhibition hall 2 central cylinder model
Main materials: 3mm laser cut Aosong board, 3mm cork roll, 5mm cork roll, white grass powder

● 效果图　Rendering

庭院透视

展馆入口空间

外廊坡道

休息空间

展厅效果图

专题六：“旧城更新”城市与建筑设计（新四校联合毕设）
Topic 6: Urban and Architectural Design of "Old City Renewal"

中山大道工艺文化街巷主入口

北侧下沉广场主入口

设计题目： 续巷行舟 ——"汉口记忆"文化体验区设计
指导老师： 严湘琦
学　　生： 蔡雨希

15

Design topic: OLD STREETS & LOCAL LIFE——"Hankou Memory" Cultural Experience Area Design

Instructor: Yan Xiangqi

Student: Cai Yuxi

● **设计说明**

设计片区主人流来向为江汉路商圈与孙中山铜像及周边商圈，具有巨大商业价值；历史建筑遗产丰富，里外生活鲜活，具有巨大历史文化价值。我们由此设置片区主轴线始终点，设计片区对外开放性（迎接外来客流），主要沿中山大道与三民路界面；对内开放性（周边社区居民），主要沿统一街界面，打铜街与统一街交汇处为对内开放性重要节点。

Design notes

The main crowds of the design area is from by the business district of Jianghan Road, the bronze statue of Sun Yat-sen and the surrounding business district, which has huge commercial value, rich historical architectural heritage, and fresh life inside and outside, with huge historical and cultural value. Therefore, we set the starting and ending points of the main axis of the area, and designed that the openness of the area to the outside world (to meet external passenger flow) is mainly along the interface of Zhongshan Avenue and Samin Road, and the openness to the inside (to residents of surrounding communities) is mainly along the interface of Tongyi Street, and the intersection of Datong Street ang Tongyi Street is an important node.

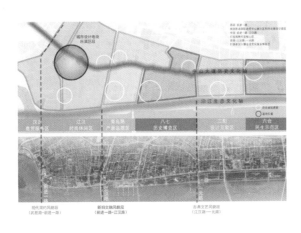

● 城市设计　Urban design

选取地块
节点

图片来源：左一，右列上数第二张由城市设计小组共同绘制，右列下部两张图片选自网络

总平面图

● 选址分析　Site selection analysis

开发自由度　　　　　　　　历史风貌分区

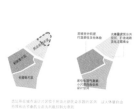

车行人行势力图　　　　　　　环境分析

选址地块开发自由度大，面向中山大道和统一街界面开发自由度差异大，面向中山大道开发自由度小，限高贴线率保障中山大道历史风貌，面向统一街开发自由度高。

The development freedom of the selected site is large, so is the development freedom difference between the interface of Zhongshan Avenue and Tongyi Street is large, and the development freedom for the interface of Zhongshan Avenue is small, and the historical style of Zhongshan Avenue is guaranteed by the line sticking rate of height limit, while the development freedom for the interface of Tongyi Street is high.

● 功能策划　Function planning

概念　　　　　　体块功能　　　　　初步任务书

● 体块生成　Block generation

人流分析

新老建筑体块关系

中山大道风貌控制——贴线率

双重开放性

新老建筑视线关系

承接江汉路商圈人流

● 高度控制　Height control

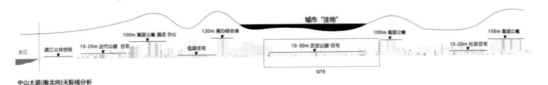

中山大道(南北向)天际线分析

● 公共空间节点设计　Public space node design

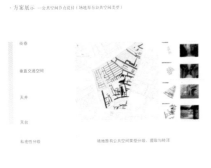

● 方案展示　Scheme display

·方案展示 —功能布局与联系

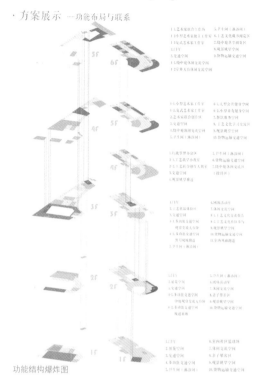
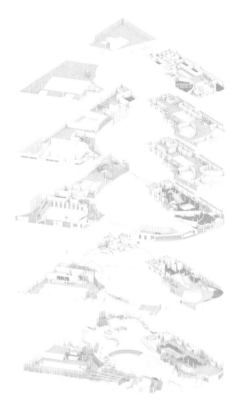

功能结构爆炸图

·方案展示 —路径体验

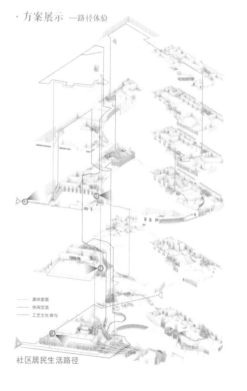

— 素饰家居
— 休闲交流
— 工艺文化参与

社区居民生活路径

① 室外健康步道—室内风雨跑道

② 社区生活文化馆—零售天台

● 效果图　Rendering

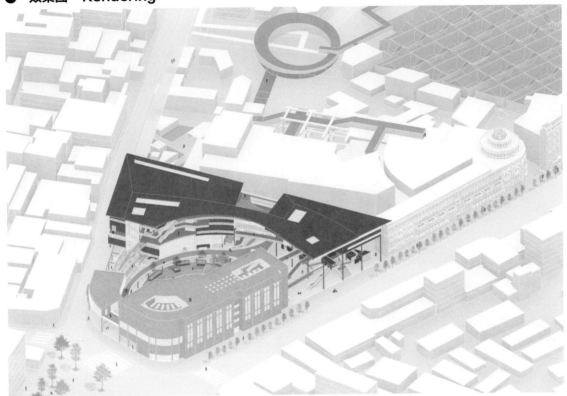

● 剖面图　Profile

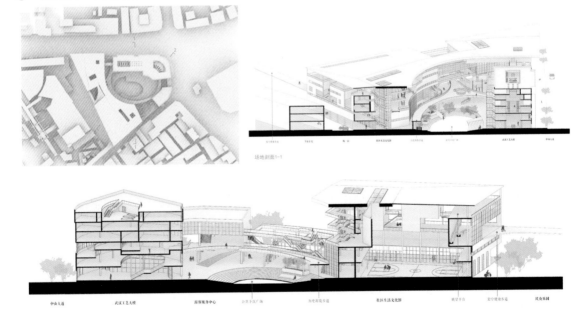

场地剖面1-1

场地剖面2-2

场地现状

鸟瞰图

设计题目： 民众乐园 | 游园复兴 ——"汉口记忆"文化体验区设计
指导老师： 陈翚
学　　生： 刘小凡

16

Design topic: People's Paradise and Garden Revival——"Hankou Memory" Cultural Experience Area Design

Instructor: Chen Hui

Student: Liu Xiaofan

● **设计说明**

正如汉口的发展与历史沿革，基地周边的地名符号也存在众多当时的地缘聚集与业缘聚集印记。

在历史地图上，几乎半数的街巷名称都与这条街上所卖的商品类别有关，也有许多与行业、寺庙、姓氏相关的街巷命名，例如药王巷、延寿庵巷、鲍家巷、田家巷，等等。

Design notes

Just like the development and historical evolution of Hankou, the place names around the base also have many marks of geographical agglomeration and industrial agglomeration at that time.

On the ancient historical map, almost half of the streets are related to the categories of goods sold on the street, and there are also many street names related to industries, temples and surnames, such as the King of Medicine Lane, Life Extension Temple Lane, Bao's Lane, Tian's Lane and so on.

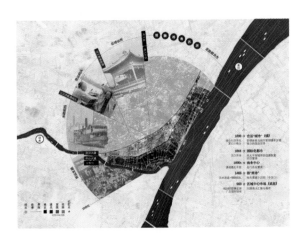

● 调研分析　Investigation analysis

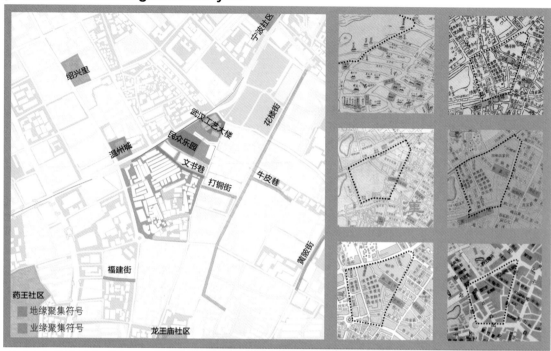

■ 美容美发业　■ 五金电器业　■ 布料纺织业
■ 电子电器业　■ 水暖灯具业　■ 大型综合商场

汉口新市场	武汉沦陷	中华人民共和国成立	改革开放
1919	1938	1949	1978
随着汉口新市场即民众乐园建成，附近各种性质的场所相继落成。该区域成为汉口华届最新潮的地区，吃喝玩乐一应俱全。	该区域成为日军控制武汉的主要场所，商业发展较迟缓。同时期，积庆里沦为慰安所。	将该区域娱乐民众乐园重新整顿，使得该区域成为更为健康的市民文化活动中心。	小商品贸易迅速发展，成为汉口重要的服贸市场和五金市场。此区域成为人们购物的首选之地。

● 概念愿景　Conceptual vision

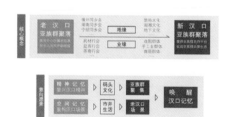

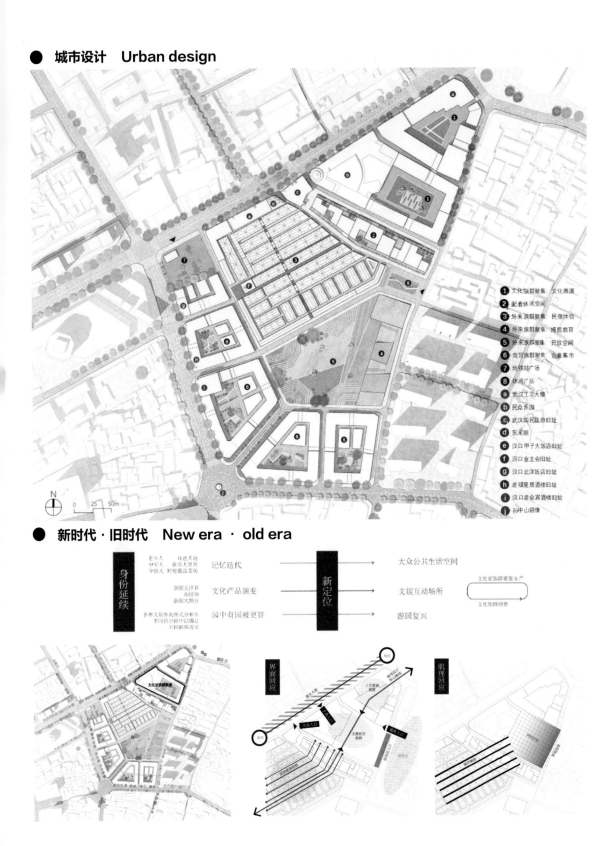

① 文化族群聚集：文化表演
② 配套休闲空间
③ 外来族群聚集：民宿体验
④ 外来族群聚集：博览教育
⑤ 外来族群聚集：开放空间
⑥ 商贸族群聚集：创意集市
⑦ 地铁站广场
⑧ 休闲广场
ⓐ 武汉工艺大楼
ⓑ 民众乐园
ⓒ 武汉国民政府旧址
ⓓ 东来顺
ⓔ 汉口甲子大饭店旧址
ⓕ 汉口业主会旧址
ⓖ 汉口北洋饭店旧址
ⓗ 老福星居酒楼旧址
ⓘ 汉口老会宾酒楼旧址
ⓙ 孙中山铜像

● 新时代 · 旧时代　New era · old era

● 城市对话建筑　Urban dialogue architecture

网格

渗透

连续

网格：积庆里民居形成的空间网格对塑造片区的尺度。
渗透：周边民居建筑形成丰富的街巷形态向场地内渗透。
连续：中山大道沿线建筑以围合或整体体块塑造街道界面。

Grid: the spatial grid formed by jiqingli folk houses is the scale of shaping the area.
Infiltration: the surrounding residential buildings form a rich form of streets and alleys and penetrate into the site.
Continuity: the buildings along Zhongshan Avenue shape the street interface with enclosure or integral blocks.

● 改造策略　Transformation strategy

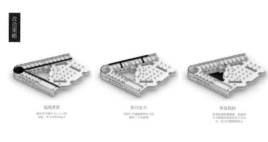
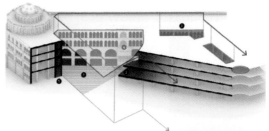

流线重置　　积行复兴　　恢复庭园

❶ 通高中庭　　　　❷ 看台　　　　❸ 内部庭院　　　　❹ 小看台　　　　❺ 内廊道

● 模度转译 · 新旧衔接　Modulus translation · Connection between old and new

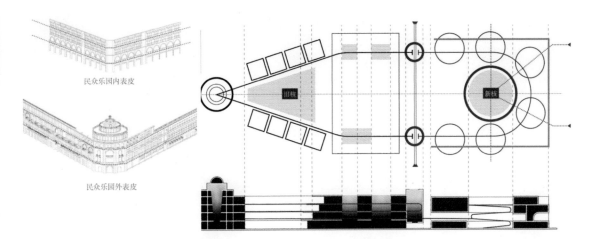

民众乐园内表皮

民众乐园外表皮

● 总平迭代　Plan iteration

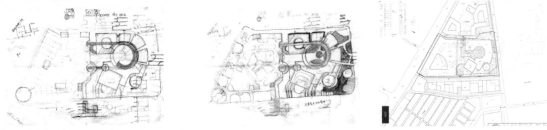

● 形体适应　Body adaptation

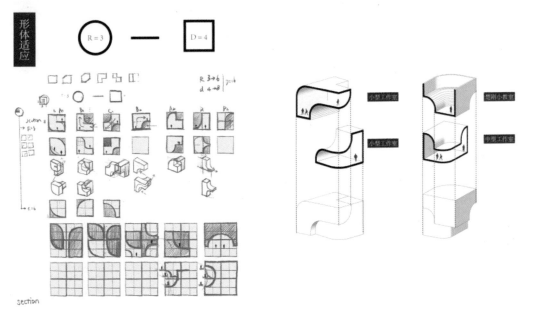

沿街界面

沿河界面

17

设计题目： 文化演绎，活力再生——四校联合毕业设计
指导老师： 严湘琦
学　　生： 徐子牧

Design topic: Cultural Interpretation, Vitality Regeneration —— The Joint Graduation Project of Four Universities

Instructor: Yan Xiangqi

Student: Xu Zimu

● 设计说明

目前用地分区分为三种：商业用地、文化设施用地及商住用地，这也和前期对于场地的分析定位有关。地面步行系统在下沉客厅处形成汇集和再发散。车行系统如图所示，新增了两条与高速北侧城市干道联通的城市道路。由于下沉广场及地下商业街的存在，地下车库设置在-2F，设置三个出入口。西南侧地块地下部分与地铁出连接，人可以通过地下街直接到达客厅。场地周边几个比较重要的节点是三个世居，场地中主要的公共空间是下沉客厅、三处内外界面联系开口，以及新增的地铁 C 出口和世居的入口广场。空中步行系统主要是串联各建筑的廊道，与建筑内的垂直交通盒子对接。

Design notes

Land use zoning is currently divided into three types: commercial land, cultural facility land and commercial and residential land, which is also related to the location of the site in the preliminary analysis. The ground walking system converges and diverges in the sunken living room. The vehicle-running system is shown in the figure, and two new urban roads connected with the northern urban trunk road of the expressway have been added. Due to the existence of the sunken plaza and underground commercial street, the underground garage is set at −2F, with three entrances and exits. The underground part of the southwest plot is connected with the subway exit, and people can reach the living room directly through the underground street. Several important nodes around the site are the three residences. The main public spaces in the site are the sunken living room, three openings of the internal and external interface, the newly added subway exit C and the entrance square of the residences. The aerial walking system mainly connects the corridors of each building and connects with the vertical traffic box inside the building.

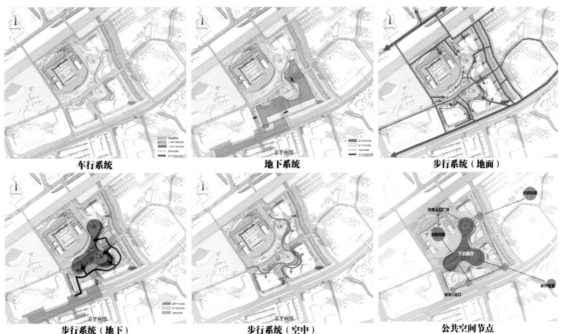

车行系统　　　　　地下系统　　　　　步行系统（地面）

步行系统（地下）　　　步行系统（空中）　　　公共空间节点

● 单体生成逻辑 Monomer generation logic

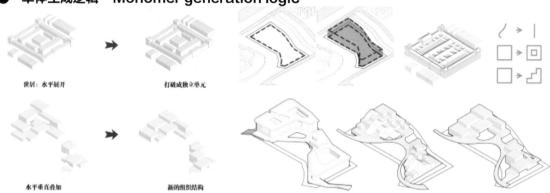

世居：水平展开　　　　　　打破成独立单元

水平垂直叠加　　　　　　新的组织结构

该地块的建筑控制线如图虚线所示。整个地块的限高为
18 m，考虑到其公建性质，将层数定在 4 层。将建筑与
世居的对话元素主要提取为三点，将城市设计较为曲线
的形式改为直线形式，平面借鉴天井元素，立面设计为
退台的形式。根据前两个失败方案的设计经验，推敲选
定了最终方案，这个方案改进了之前方案的不足，底层
局部架空不仅让采光通风问题得到了解决，还形成了开
放的广场，将紧张的空间还给城市，同时实现了良好的
流线可达性。将零散的体块整合，满足了之前退台的需求，
仍然保留了世居背山靠水的意向。

该方案的生成逻辑可以理解为，世居是水平展开的多种
独立单元，将世居解构打破，再水平和垂直叠加，形成
新的组织结构。

The building restriction line of the plot is shown as the dotted line.
The height limit of the whole plot is 18 meters. Considering its nature
of public construction, I set the number of floors at four. I extracted
the dialogue elements between architecture and residence mainly
into three points, and changed the relatively curved form of urban
design into a straight form. The plane used patio elements for
reference, and the facade design was in the form of terrace. Based
on the failed design experience of the first two schemes, I carefully
selected the final scheme, which improved the shortcomings of
the previous schemes. The partial overhead at the bottom not only
solved the problem of lighting and ventilation, but also formed an
open square, returning the tight space to the city and realizing
good circulation accessibility. The scattered blocks are integrated
to meet the needs of the previous terrace, but also still retain the
residents' intention of living besides the mountains with water.

The generating logic of the scheme can be understood as that the
house is a variety of independent units that are unfolded horizontally.
The house is deconstructed and then superposed horizontally and
vertically to form a new organizational structure.

● 场地设计策略　Site design strategy

置入建筑体块　　　　设计廊道串联地块与建筑

人流引导向北入口　　城道将人群由下沉广场引向地面

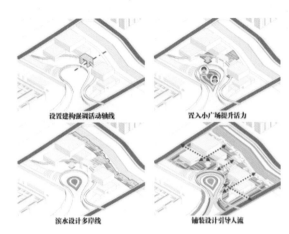

设置建构强调活动轴线　　置入小广场提升活力

滨水设计多岸线　　　　铺装设计引导人流

设置建构强调中心活动轴线，置入小广场提升场地活力，滨水设计不同标高的多岸线，使游玩的人们在不同水位期都能与水体近距离接触。场地铺装设计可以引导人流通过各个架空节点。

Set and construct the axis that emphasize central activities, place a small square to enhance the vitality of the site, and design multiple shorelines with different elevations along the waterfront, so that people can have close contact with the water body at different water levels. The pavement design of the site can guide the flow of people through each overhead node.

● 功能流线　Functional streamline

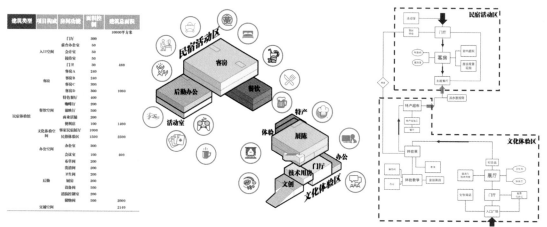

功能分区大致将建筑分为南北两馆，南馆主要为文化体验区，北馆为民宿活动区。将餐饮与体验部分放置沿河处，有利于与观景体系的结合。活动与商业功能放置在一、二层保证其可达性。办公区域放置在主要道路旁，方便后勤人员出入。展陈与客房功能放置在楼层较高的位置，保证其私密性。

具体的活动流线设置：南馆由门厅上至三、四层展厅，再下至二层的体验区和一层的特产超市，由次入口到达滨水景观带。北馆由门厅接受前台接待，并上至三、四层客房。一层配有桌游、麻将等活动室，客房配有室内庭院与屋顶观景花园，客房直接联通去往二楼的餐厅，并可由次入口出入滨水景观带。

The functional zoning roughly divides the building into two pavilions: the south pavilion is mainly for cultural experience, and the north pavilion is for homestay activities. The dining and experience section is placed along the river to facilitate the integration with the viewing system. Activities and business functions are placed on the first and second floors to ensure their accessibility. The office area is placed next to the main road to facilitate the access of support personnel. Exhibits and guest rooms are placed on higher floors to ensure their privacy.

The specific activity flow is set as follows: the south pavilion goes up from the hallway to the exhibition hall on the third and fourth floors, then down to the experience area on the second floor and the specialty supermarket on the first floor, and from the secondary entrance to the waterfront landscape belt. North pavilion from the lobby reception, and up to the third and fourth floor guest rooms. The first floor is equipped with an activity room for table games and mahjong, and the guest room is equipped with an indoor courtyard and rooftop viewing garden. The guest room is directly connected to the restaurant on the second floor, and the waterfront viewing belt can be accessed from the secondary entrance.

● 界面处理　Interface processing

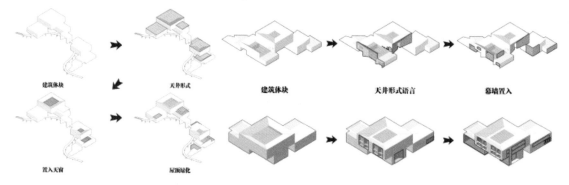

屋顶界面对世居天井元素进行了借鉴与转译，置入天窗采光，并设计了屋顶绿化与活动空间。这是屋顶大样与屋顶排水设计的平面图，采用天沟内排水的形式。立面也同样采取了天井的形式语言，并与玻璃幕墙相结合。

First of all, the elements of the residence patio is used for reference and translation of the roof interface and the skylight is inserted for lighting, and the roof greening and activity space is designed. This is the size of the roof, and the floor plan of the roof drainage design, using the form of drainage in the gutter. The facade also adopts the formal language of the patio and is integrated with the glass curtain wall.

● 内部空间提炼　Internal space refining

内部空间的具体设计手法提炼了如下几种世居的空间原型：首先提取了世居门厅的空间原型。世居的门厅作为一个高峨的入口空间，转译为展厅的门厅和民宿的大堂，采用两层通高的形式，营造进入建筑的空间仪式感。这是展馆门厅的效果图，我们也可以从一、二层平面图看到这两个通高的空间。

其次提取世居祠堂作为原型。祠堂位于世居的核心位置，是家族精神寄托的中心。我们将其转译为南馆的展厅和北馆的客房，它们分别发挥两栋建筑最核心的功能，在这里人员的活动与交流发生的频率很高，并且采用一个完整的天井屋顶形式，天井处集中天窗采光，形成空间的神圣感。

展厅中央设计为天窗采光的通高展厅，作为观展流线的高潮，还设计了可以直接与世居对话的观景厅。
北馆的客房区也在天井处设置了室内庭院，每个客房皆面向且直通庭院，传承了世居居住的共享体验，增进了住客的情感交流。

The specific design method of the interior space abstracts the following space prototypes from the residences. Firstly, the space prototype of the hall of the residence is extracted. The hall of the residence, as a lofty entrance space, is translated into the hall of the exhibition hall and the hall of the homestay, which adopts the form of double-height, creating a sense of space ritual into the building. This is a rendering of the foyer of the pavilion. We can also see these two double-height spaces from the floor plan.

Secondly, ancestral hall is extracted as the prototype. Ancestral hall is located at the core of the ancestral residence and is the center of the spiritual sustenance of the family. We translate it into the exhibition hall in the south hall and the guest room in the north Hall, which are the core functions of the two buildings respectively. Here, people's activities and exchanges take place in a high frequency. Moreover, a complete roof form of the courtyard is adopted, where skylight lighting is concentrated, forming a sense of holiness in the space.

The center of the exhibition hall is designed as a double-height exhibition hall with skylight lighting, as the climax of the exhibition flow, and also designed a viewing hall that can directly communicate with the residents of the world.

The guest room area of the north pavilion also has an indoor courtyard at the patio. Each guest room faces and leads directly to the courtyard, inheriting the shared experience of living in the world and enhancing the emotional communication of the guests.

· 内部空间原型提炼

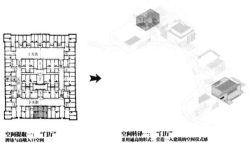

空间提取一："门厅"
牌场与高峨入口空间

空间转译一："门厅"
采用通高的形式，营造一入建筑的空间仪式感

· 内部空间原型提炼

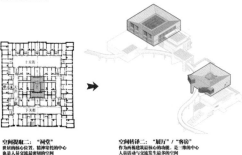

空间提取二："祠堂"
世居的核心位置，精神寄托的中心
也是人员交流最密切的空间

空间转译二："展厅"/"客房"
作为两栋建筑最核心的功能，是二维的中心
人员活动与交流发生最多的空间
完整的天井形式

对江书座

活动广场

设计题目：安化县黑茶文化体验中心
指导老师：严湘琦 陈翠 罗莨
学　　生：周婷

18

Design topic: Anhua Dark Tea Cultural Experience Center

Instructors: Yan Xiangqi, Chen Hui, Luo Jin

Student: Zhou Ting

● 设计说明

本设计致力于打造高端茶旅一体化设计：营造活化滨水的公共空间，戛然而止的古街，趋于光明的码头，以历史建筑为展。以人为本，关注原住民物质与精神需求，顺应城市生态体系建设，保护历史文脉，尊重历史建筑，解析城市历史文化因素，反映历史文脉，反映时代特色，以高起点的环境艺术及景观设计创造一个现代化的展示与体验中心，让其成为城市的特色名片，成为城市的历史文化载体。

Design notes

This design is committed to creating a high-end integrated design of tea travel: creating an activated public space along the waterfront, an ancient street that stops abruptly, a pier that tends to be bright, and a historical building as an exhibition. Take people as our priority, focus on indigenous people's material and spiritual needs, adapt to urban ecological system, protect historical context, respect for historical buildings, analyze urban historical and cultural factors, reflect the historical context, reflect the era characteristics, with a high starting point of environment art and landscape design to create a modern exhibition and experience center, make it become the characteristics of city business card, become the historical and cultural carrier of the city.

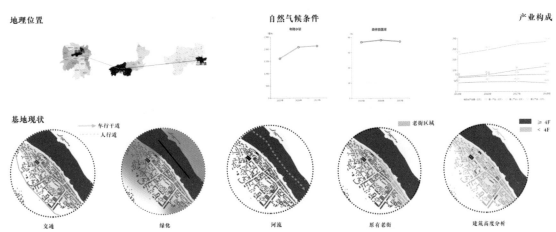

地理位置　　　　自然气候条件　　　　产业构成

基地现状　　→ 车行干道　　----> 人行道　　老街区域　　≥ 4F　　< 4F

交通　　　　绿化　　　　河流　　　　原有老街　　　　建筑高度分析

● **江南镇空间历史演变　Spatial historical evolution of Jiangnan town**

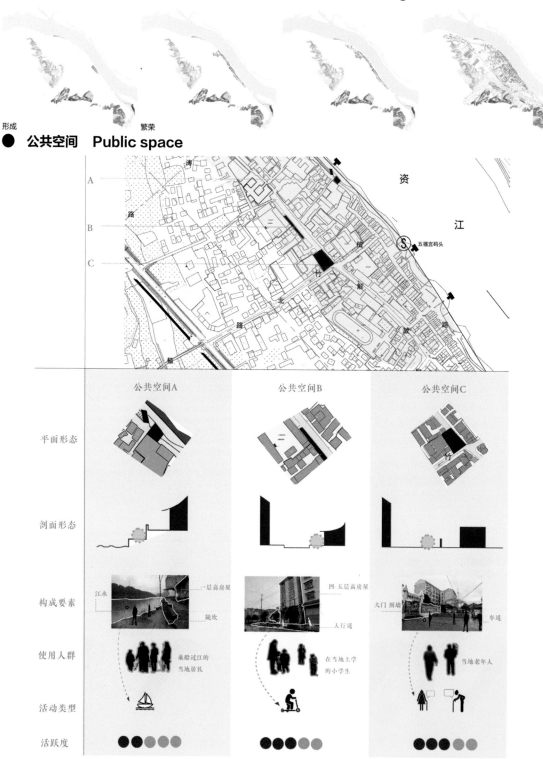

形成　　　　　　　　　　　　繁荣

● **公共空间　Public space**

● 图底肌理　Figure bottom texture

紧密围合空间	围合较强空间	围合较弱空间	散布空间

图底肌理关系分析

江南镇主要有两种肌理关系:

一是通过人为规划、设计的肌理,如现代公共建筑、商业街等。

二是居民自发建造的建筑形成的肌理,主要的肌理关系是通过建筑自然围合成公共空间,通过围合感的强弱可分为左边四种类型。其共同特点是空间完形程度低,形状自由,无明显轴线。

● 选址分析　Site selection analysis

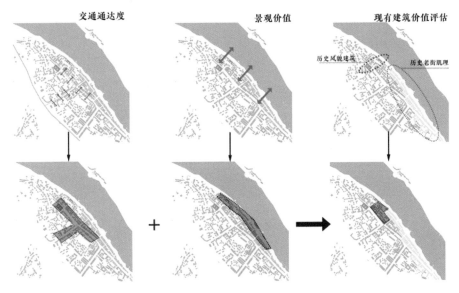

交通通达度　　　　景观价值　　　　现有建筑价值评估

历史风貌建筑　　历史老街肌理

● 场地分析　Site analysis

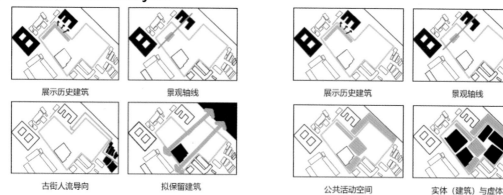

展示历史建筑　景观轴线

古街人流导向　拟保留建筑

展示历史建筑　景观轴线

公共活动空间　实体（建筑）与虚体

现有保护建筑三处，其中五福宫码头主要用于古代茶叶运输，在功能上与建筑不便有较大联系，综合考虑功能与观赏修缮价值，选择靠近老建筑处。

There are three protection buildings, among which wufu Palace wharf is mainly used for ancient tea transportation, which has a great connection with the inconvenience of the building. Considering the function and the value of ornamental repair, it is selected to be close to the old building.

● 体块生成　Block generation

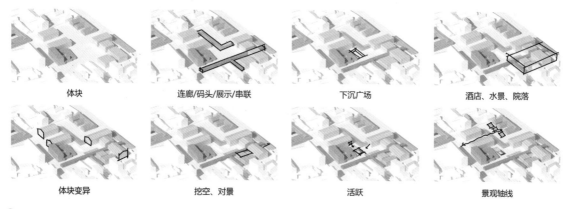

体块　连廊/码头/展示/串联　下沉广场　酒店、水景、院落

体块变异　挖空、对景　活跃　景观轴线

● 总平面图　General layout

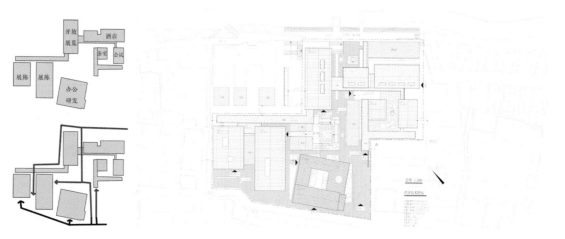

● 码头空间意象　Wharf space intention

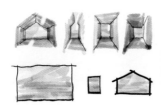

缘溪行，忘路之远近。忽逢桃花林，夹岸数百步，中无杂树，芳草鲜美，落英缤纷。渔人甚异之，复前行，欲穷其林。林尽水源，便得一山，山有小口，仿佛若有光。便舍船，从口入。初极狭，才通人。复行数十步，豁然开朗。

—————陶渊明《桃花源记》

Walking along a streamline, with no sense of how far away from the starting point. All of a sudden, a peach-blossom-grove was founded, and on both sides across the streamline with hundreds-of-feet-wide land, no other trees but peach ones grow with fresh and sweet grass petals falling in riotous profusion. Pass through the whole grove reaching a fountainhead near a mountain. There was a small hole seemingly glittering in front of the mountain. Leaving the boat, enter the hole as an open door being so narrow that only one man could press himself to penetrate into it. Continuing for over tens-of-feet distance, the view instantly clear up.

-- Tao Yuanming, *The Land of Peach Blossoms*

● 效果图　Rendering

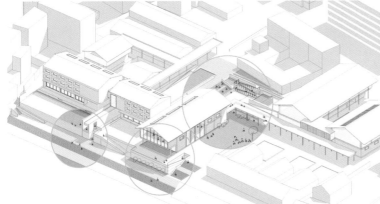

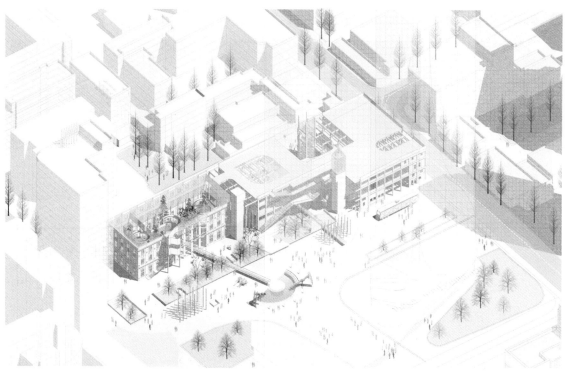

鸟瞰图

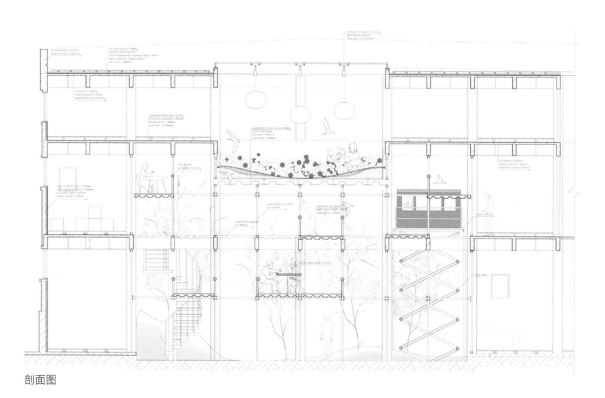

剖面图

19

设计题目： 奇境乐园
指导老师： 张蔚 谢菲
学　　生： 刘恬

Design topic: Wonderland Park

Instructors: Zhang Wei, Xie Fei

Student: Liu Tian

● **设计说明**

旧汉口存在着一种繁荣的状态，它曾是个露天的市场。林立的茶馆中，武汉旧生活依然热气腾腾：卖唱的、说书的、湖北大鼓的，还有穿黑衣、涂白脸、挎着竹篮子的唱婆子穿梭其间，像庙会一样热闹。

如今的汉口却是十分冷静，沿着一元路胜利街走过去，空间平铺直叙，直白得不近人情，租界没有被刻在历史的耻辱柱上，而是与市民生活长久的剥离开来，偶尔经过短小的里弄，还可感受到旧汉口的风韵和气息。

Design notes

There is a flourishing state in old Hankou. It used to be an open-air market. Among the teahouses, the old life of Wuhan is still busy.

Now Hankou is somewhat deserted, walking along the Victory Street of One yuan Road, the space is flat and straight, straightforward and inhuman, the concession has not been carved on the historical humiliation column, but it has been separated from the civic life for a long time, only occasionally through the short lane, one can feel some of the charm and flavor of the old Hankou.

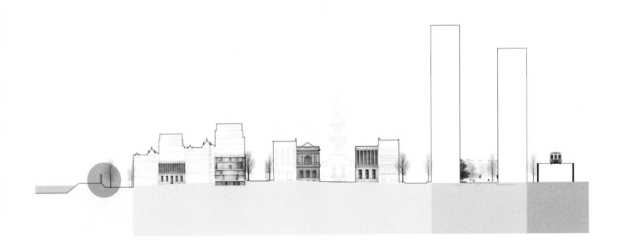

● 租界 1920-1980　Concession 1920-1980

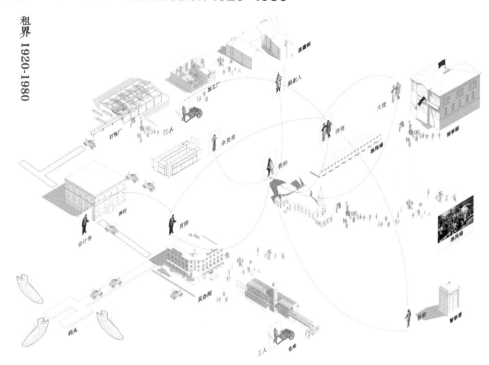

● 租界系统转译　Concession system translation

商业

娱乐

生活

● **界面改造   Interface redesign**

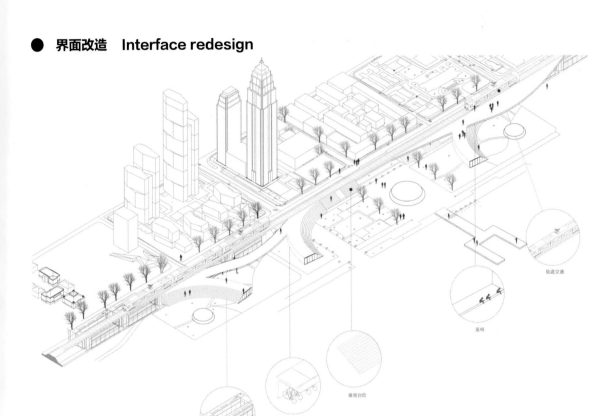

轨道交通

座椅

景观台阶

自行车停车

● **城市设计概念   Urban design concept**

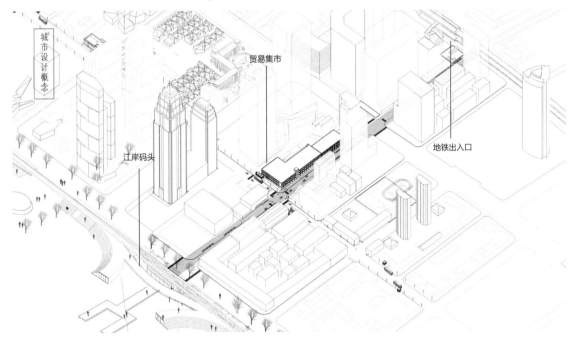

城市设计概念

贸易集市

地铁出入口

江岸码头

● 生成过程　Generation process

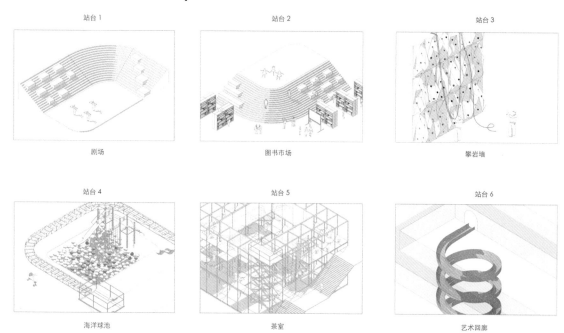

站台 1

剧场

站台 2

图书市场

站台 3

攀岩墙

站台 4

海洋球池

站台 5

茶室

站台 6

艺术回廊

● 解构图　Deconstruction diagram

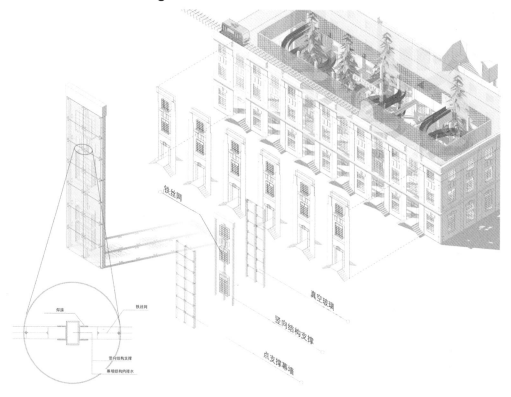

铁丝网

真空玻璃

竖向结构支撑

点支撑幕墙

焊接

铁丝网

竖向结构支撑

幕墙结构内排水

● 室内效果　Indoor effect

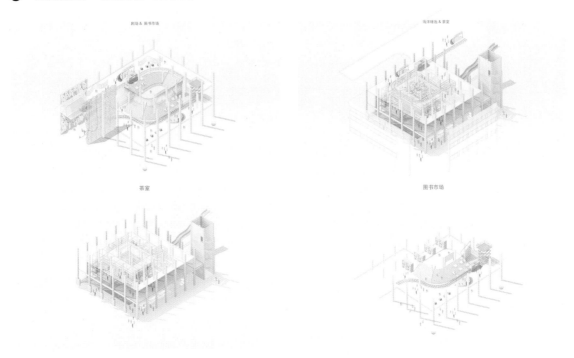

● 整体鸟瞰　Overall aerial view

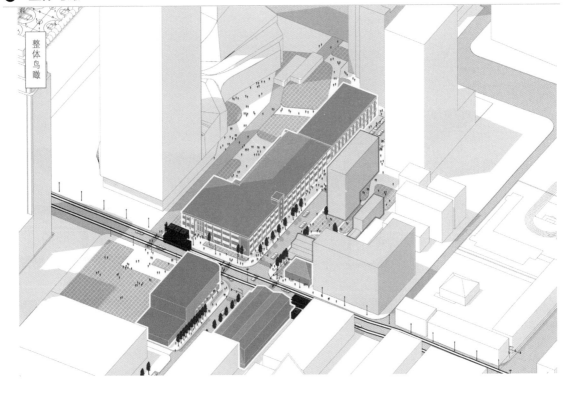

透视效果图

剖面图

设计题目： 观游山海——大连理工大学艺术博物馆设计
指导老师： 宋明星 卢健松
学　　生： 莫杨晨露

20

Design topic: Visiting Mountains and Seas —— Design of Art Museum of Dalian University of Technology

Instructors: Song Mingxing, Lu Jiansong

Student: Moyang Chenlu

● **设计说明**

项目北面为城市主干道，保留建筑临街立面原本的特色，能够吸引大量市民进入场地。南侧也为城市主干道，且有多条支路交汇于此，能够形成城市广场空间。南北新老对比作为本项目考虑的一个重要元素，同时用一个完整的界面将场地中的多种功能复合串联。建筑布局为一个博物馆 L 形组团，和两个高层办公点状布局，建筑充分利用场地条件，力求组织布局合理，功能分区明确。人们行走在场地中能感受市民空间的自由，增加人与人之间的对话和交流。

Design notes

To the north of the project is the main street of the city, preserving the distinctive street facade of the building to attract a large number of citizens into the site. The south side is also the main road of the city, and there are several branches where the intersection can form the space of the city square. The contrast between the old and the new between the north and the south is an important element of the project, and a complete interface is used to connect various functions in the site. The layout of the building is an L-shaped group of museums and two high-rise office points. The building makes full use of the site conditions, and strives to make the layout reasonable and the functional zoning clear. People walking on the site can feel the freedom of the civic space, increasing the dialogue and communication between people.

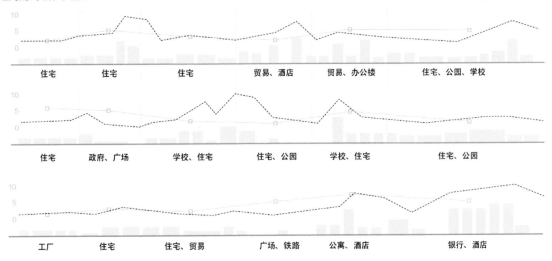

● 城市研究 Urban research

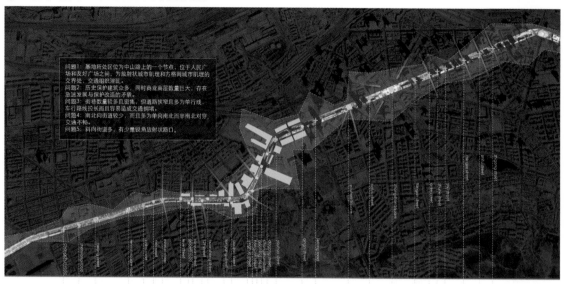

问题1：基地所处区位为中山路上的一个节点，位于人民广场和友好广场之间，为放射状城市肌理和方格网城市肌理的交界处、交通组织混乱。
问题2：历史保护建筑众多，同时商业高层数量巨大，存在急速发展和保护改造的矛盾。
问题3：街巷数量较多且密集，但道路狭窄且多为单行线，车行路线拉长而且容易造成交通拥堵。
问题4：南北向街道较少，而且多为单向南北而非南北对穿，交通不畅。
问题5：斜向街道多，有少量锐角放射状路口。

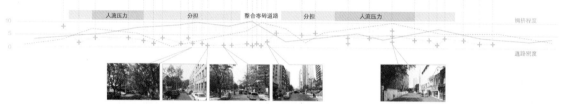

大连理工大学市内校区位于西岗区一二九街、中山路两侧，青泥洼商圈和人民广场之间，毗邻劳动公园。占地面积约 9.45hm²，占有良好的地理优势和土地资源。自 2010 年大连理工大学西部校区正式启用，大连理工大学化工学院回迁学校本部，市内化工学院至今一直处于闲置的状态。

Dalian University of Technology (DUT) campus is located at 129th Street, Xigang District, on both sides of Zhongshan Road, between Qingniwa Business Circle and People's Square, adjacent to labor Park. Covering an area of about 9.45 hectares, it has a good geographical advantages and land resources. Since 2010, when the western campus of Dalian University of Technology (DUT) was officially opened and the college of Chemical Engineering of DUT moved back to its main campus, the college of Chemical Engineering in the city has been idle until now.

● 概念生成 Concept generation

● 流线分析　Streamline analysis

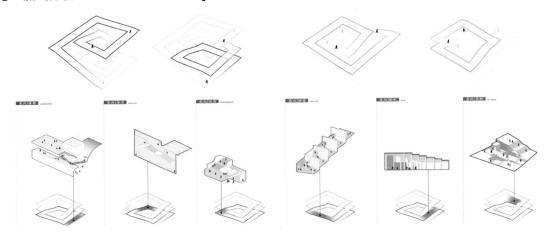

● 总平面图　General layout

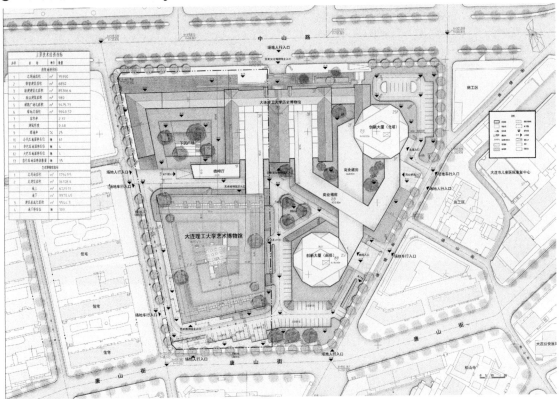

保留北侧具有文化价值的日字形建筑及其西侧的 L 形建筑，将其改造成为大连理工大学的人文历史博物馆，西南侧作为延续，新建一座大连理工大学的艺术博物馆，两馆中间以一个下沉广场作为连接。

The Chinese character shaped building with cultural value on the north side and the L-shaped building on the west side will be retained and transformed into the humanities and history museum of Dalian University of Technology. The art museum of Dalian University of Technology will be built on the southwest side as a continuation. The two museums will be connected by a sunken square in the middle.

● 平面图　Plan

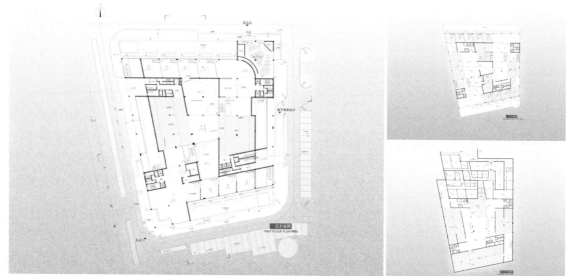

建筑一层主要功能为临时展厅、门厅、咖啡厅、报告厅，通过阶梯与负一层相连。二层主设两个大展厅及一个小展厅、一个放映厅，并设有两个开放式阶梯，具有复合展陈、休息、交通多种功能。地下一层北侧设有办公、藏品管理、设备用房；南侧为四个大型展厅，空间为二层通高，同时设有高侧窗引入光线，满足采光需求。地下二层为车库，设有两个出入口，共提供 109 个机动车位。

The main functions of the first floor of the building are temporary exhibition hall, hallway, cafe and lecture hall, which are connected to the negative first floor by stairs. There are two large exhibition halls, one small exhibition hall and one screening hall on the second floor, and there are two open stairs for composite exhibition, rest and transportation. The north side of the first underground floor is equipped with office, collection management, equipment room; There are four large exhibition halls on the south side. The space is double-height on the second floor, and there are high side window to introduce light to meet the needs of lighting. The underground second floor is an underground garage with two entrances and exits, providing a total of 109 motor vehicle parking spaces.

● 立面图　Elevation

● 剖面图　Profile

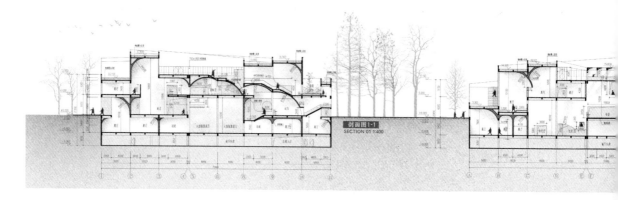

● 效果图　Rendering

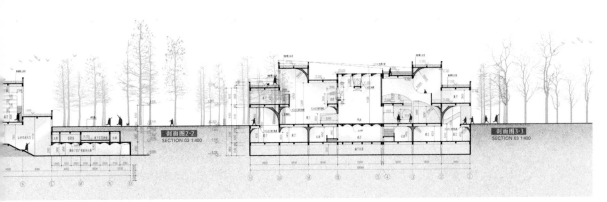

城乡规划专业优秀毕业设计

专题一：湘西怀化溆浦县城市设计
Topic 1: Land Spatial Planning and Overall Urban Design

总平面图（溆浦县总体城市设计）

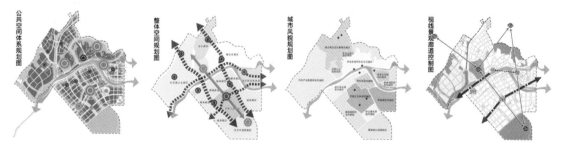

设计题目：湘西怀化溆浦县总体城市设计
指导老师：许乙青
学　　生：马文婧

1

Design topic: Overall Urban Design of Xupu County in Huaihua,Western Hunan

Instructor: Xu Yiqing

Student: Ma Wenjing

● **设计说明**

为充分结合湖南省和怀化市新定位、新战略，规划依托溆浦交通区位、产业和资源优势，溆浦县城空间发展战略规划提出 "一区一核一城"的发展战略，主动融入新一轮的湖南大发展和怀化大提升。

（1）一区（湘西示范区）：长株潭向大湘西加速发展的过渡示范区；（2）一核（怀化增长核）：融入怀化 "一极两带"的东北部重要增长核；（3）一城（溆水中心城）：带动溆水流域全方位发展的生态旅游和商贸物流中心城。

Design notes

In order to fully combine the new positioning and new strategy of Hunan Province and Huaihua City, and rely on the advantages of traffic location, industry and resources, the spatial development strategic plan of Xupu County puts forward the development strategy of "one area, one core and one city"to actively integrate into the new round of Hunan development and Huaihua promotion.

(1) One area (Xiangxi demonstration area): the transition demonstration area between Chang-Zhu-Tan and western Hunan; (2) One core (Huaihua growth core): an important growth core in the Northeast integrated into the "one pole and two belts" of Huaihua; (3) One city (Xushui central city): it is an eco-tourism and trade logistics center city that drives the all-round development of Xushui River Basin.

125

总平面图（溆浦县总体城市设计）

鸟瞰图（溆浦县总体城市设计）

设计题目：湘西怀化溆浦县总体城市设计
指导老师：许乙青
学　　生：张宁致

Design topic: Overall Urban Design of Xupu County in Huaihua,Western Hunan
Instructor: Xu Yiqing
Student: Zhang Ningzhi

● **设计说明**

本方案以"山水田园、城乡共生"为设计理念，对溆浦县城中心城区的总体空间结构、视线景观廊道、公共空间体系、城市总体风貌、自然山水格局等方面的内容进行了规划设计，希望以景观农田串联城市总体景观结构的方式来整合城乡发展空间，完善溆浦县城"生态宜居城市"的城市性质。

Design notes

Based on the design concept of "countryside, symbiosis between urban and rural areas", this plan with natural landscape designed the overall spatial structure , sight corridors, public space system, overall urban style, natural landscape pattern of the central urban area of Xupu County, etc. , hoping to integrate urban and rural development space by connecting landscape farmland with the overall urban landscape structure, and strengthen its identity as an "ecological and livable city".

公共空间体系规划　　　　　　　　总体空间结构规划　　　　　　　　自然山水格局规划

视线景观廊道规划　　　　　　　　重点地区划分管控　　　　　　　　城市总体风貌规划

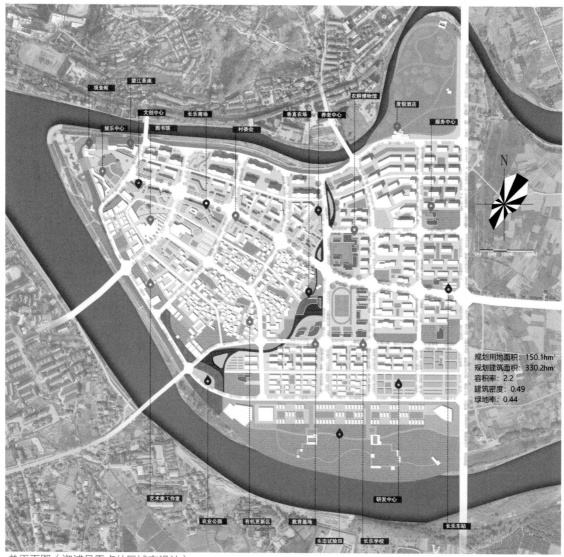

观音阁　望江茶庄　　　　　　　　　　　　　农桥博物馆
　　　　文创中心　长乐商场　　　　垂直农场　养老中心　度假酒店
娱乐中心　图书馆　　　　村委会　　　　　　　　　　服务中心

N

0M　50M　100M　200M

规划用地面积：150.1hm²
规划建筑面积：330.2hm²
容积率：2.2
建筑密度：0.49
绿地率：0.44

　　　　　　　　　　　　　　　　　　　　　研发中心
艺术家工作室
　　农业公园　有机更新区　教育基地　　　　　　　　　长乐车站
　　　　　　　　　　　　生态试验田　长乐学校

总平面图（溆浦县重点片区城市设计）

鸟瞰图（溆浦县重点片区城市设计）

设计题目：湘西怀化溆浦县长乐坊重点片区城市设计
指导老师：许乙青
学　　生：张宁致

Design topic: Urban Design of Changlefang Key Area in Xupu County in Huaihua City, Western Hunan
Instructor: Xu Yiqing
Student: Zhang Ningzhi

● 设计说明

本方案将场地内原有的耕作农田转化为具有城市景观绿地性质的城市景观农田，希望将农田的农业景观风貌与复杂的城市功能和活动融合，塑造具有田园风貌的现代城市，平衡城乡发展之间的关系。

Design notes

This plan transforms the original farmland in the site into urban landscape farmland with the nature of urban landscape green space, hoping to integrate the agricultural landscape of the farmland with complex urban functions and activities, so as to shape a modern city with an idyllic style and balance the relationship between development of urban and rural areas.

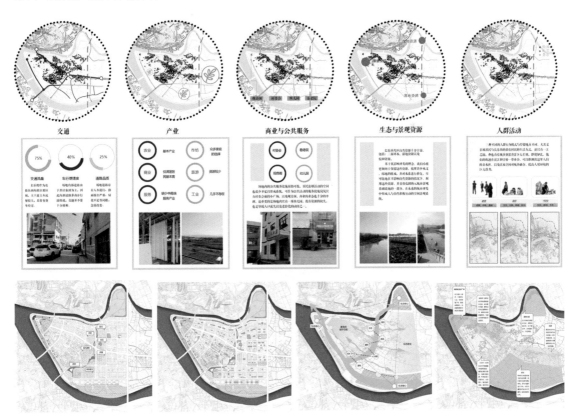

总平面图（长乐坊重点片区城市设计）

鸟瞰图（长乐坊重点片区城市设计）

设计题目：湘西怀化溆浦县长乐坊重点片区城市设计
指导老师：许乙青
学　　生：李元博

Design topic: Urban Design of Changlefang Key Area in Xupu County in Huaihua City, Western Hunan

Instructor: Xu Yiqing

Student: Li Yuanbo

4

● **设计说明**

根据现状以及城市需求的分析，确定长乐坊片区文化休闲中心的定位，同时引出对长乐坊片区城市设计的核心概念————"城市客厅"。

"城市客厅"：把这个城市的历史、发展、未来、文化的东西都综合展现在某些特别的人文景观之上，起到文化休闲的功能，是一个城市文化和形象的代表。长乐坊片区内根据政府经济能力以及现状，保留一部分老旧建筑，加以功能置换，同时拆除地方新建的公共建筑，与部分保留的老建筑共同形成长乐坊片区的核心区域————"城市客厅"。

Design notes

According to the analysis of the current situation and urban needs, the positioning of Changlefang area as the cultural and leisure center is determined. At the same time, it leads to the core concept of the urban design of Changlefang area- "living room of the city".

"Living Room of the city": It is a comprehensive display of the history, development, future, and culture of this city on some special humanistic landscape. It plays the function of culture and leisure and represent a city's culture and image. According to the government's economic capacity and the current situation, the Changlefang area retains some of the existing old buildings and replaces their functions. At the same time, the demolition of the newly built public buildings and some of the retained old buildings together form the core area of the Changlefang area-the "living room of the city".

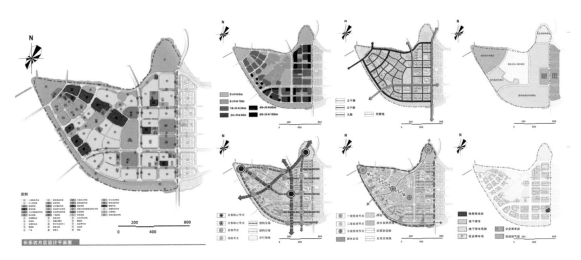

长乐坊片区设计平面图

专题二：特色主题城市设计
Topic 2: Urban Design with Specific Theme

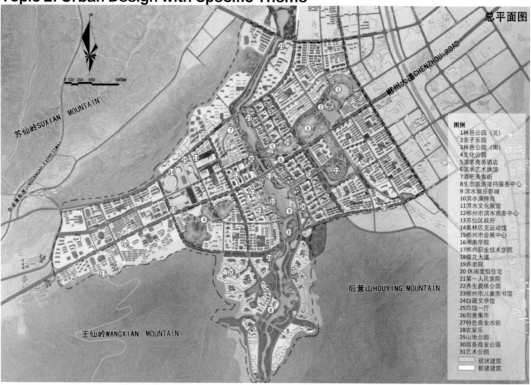

图例
1 林邑公园（北）
2 亲子乐园
3 林邑公园（南）
4 文化公园
5 滨水商务酒店
6 滨水艺术展馆
7 酒吧美食街
8 生态旅游接待服务中心
9 滨水娱乐影城
10 滨水演绎岛
11 滨水文化展馆
12 郴州市滨水商务中心
13 苏仙区政府
14 奥林匹克运动馆
15 郴州市会展中心
16 湘南学院
17 郴州职业技术学院
18 樱花大道
19 老老院
20 休闲度假住宅
21 第一人民医院
22 养生晨练公园
23 郴州市儿童图书馆
24 白薇文学馆
25 四馆一厅
26 创意集市
27 特色商业水街
28 农家乐
29 山地公园
30 商务商业公园
31 艺术公园
　现状建筑
　新建建筑

总平面图（郴州市山水城市总体城市设计）

鸟瞰图（郴州市山水城市总体城市设计）

132

设计题目：郴州市山水城市总体城市设计
指导老师：许乙青
学　　生：刘诗琪

Design topic: Overall Urban Design of Chenzhou City
Instructor: Xu Yiqing
Student: Liu Shiqi

● **设计说明**

从城市整体空间层面明确城市发展定位，提出中心城区山水城市设计总体架构框架和总体控制引导；同时在片区层面对影响城市风貌的重要因素进行研究，提出引导与控制的具体手段，增强规划的可操作性，为下阶段的控规工作提供空间上的指引，以求更好地塑造城市特色。

Design notes

Define the urban development orientation from the overall urban space level, and put forward the overall framework and overall control guidance of landscape urban design in the central urban area; At the same time, the important elements affecting the urban style are studied at the regional level, and the specific means of guidance and control are put forward to enhance the operability of the planning, so as to provide spatial guidance for the planning control work in the next stage, and then better shape the urban characteristics.

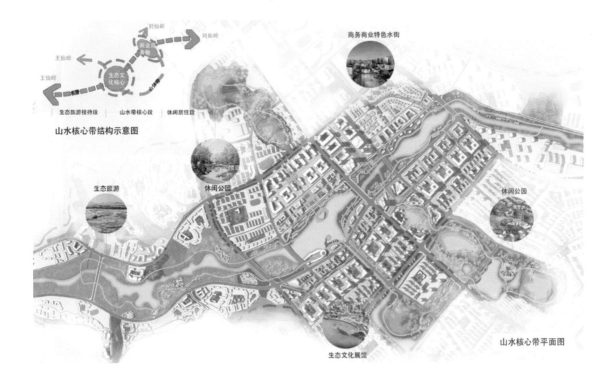

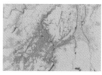

2009 年绿地系统，绿地与城市建设用地相融。

2012 年绿地系统，中心城区边缘与绿地衔接处部分被吞噬，主要为房地产开发。

2014 年绿地系统，大面积边缘绿地被吞噬，部分中心城区内公园绿地被占用，被吞噬绿地主要为道路等基础设施建设、房地产开发。两处面状水系减少。

2016 年绿地系统（现状），中心城区边缘成片的绿地被割裂成几个独立斑块，绿地系统完整性被破坏。

中心城区现状人均公共绿地偏低，低于国家园林城市8㎡/人的标准。
城区内缺乏林荫步道、街头游园等多层次绿化休闲空间，新建道路沿路绿化比重较低，道路景观单调。
老城区内建筑密度高，绿化水平偏低，居住区级公共绿地严重缺乏，绿地布局不合理，分布不均。

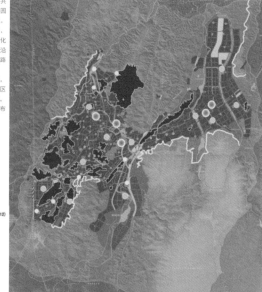

⊚　区域性综合公园（G112）
◼　居住区公园（G121）
◆　小区游园（G122）
◆　专类公园（G13）
◆　街旁绿地（G15）

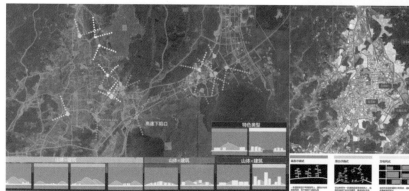

高程重分类

坡度重分类

水体分布重分类

制高点重分类

农林绿地重分类

观景点评分

岩石稳定性重分类

流域分析

矿产资源重分类

土壤侵蚀重分类

洪水淹没重分类

历史遗迹重分类

● 分析框架　Framework of the analysis

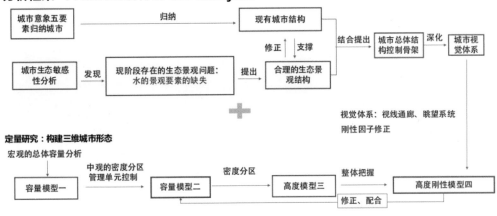

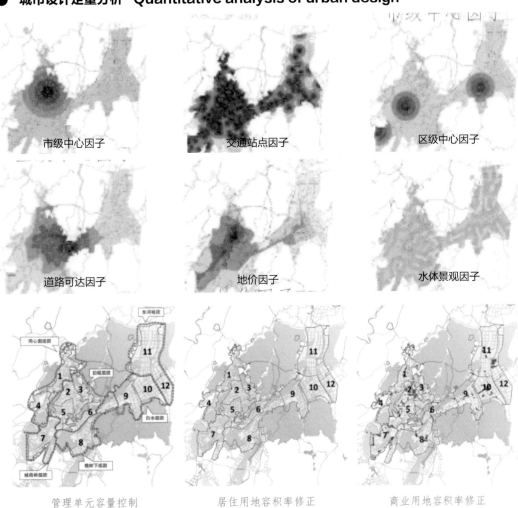

● 城市设计定量分析　Quantitative analysis of urban design

市级中心因子　　　　交通站点因子　　　　区级中心因子

道路可达因子　　　　地价因子　　　　水体景观因子

管理单元容量控制　　　居住用地容积率修正　　　商业用地容积率修正

● 城市设计控制体系 Urban design control system

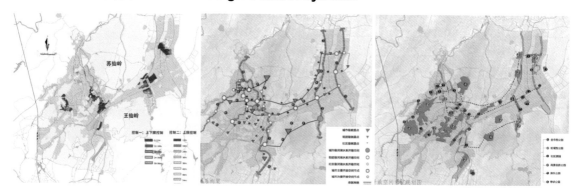

● 容积率管控分析 Plot ratio control analysis

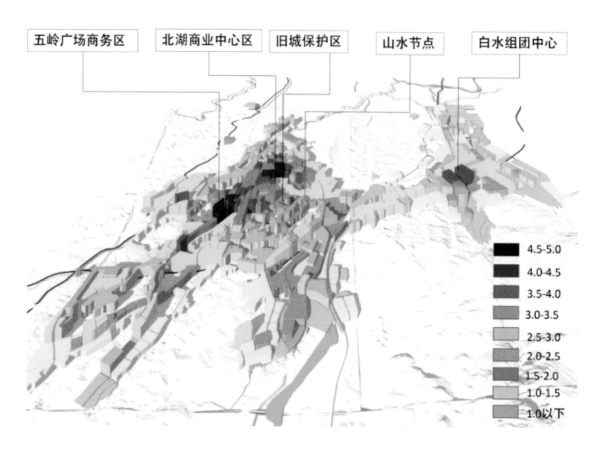

规划将组团和管理单元细分，确定每个管理单元的容积率管控范围，较客观地反映地块容积率的情况。同时，容积率数值被量化为九个等级，组团容积率控制特征显而易见，即形成了多中心结构。

The plan subdivides the Clusters and Management Units, determines the Control Range of Plot Ratio of each Management Unit, and objectively reflects the Plot Ratio of the Plot. At the same time, the Plot Ratio is quantified into nine levels, and the Control Characteristics of Cluster Plot Ratio are obvious, that is, a Polycentric Structure is formed.

● 方案表达 Scheme expression

STEP.1 梳理内部水系
基于集水流域分析，将白水组团内现有水系按照两个集水流域，依托栋溪河打造城市内部的中央水轴；其次将林邑公园附近的水系联合起来，打造城市重要的水体斑块，成为休闲居住核心。管控中，更有针对性，比较好地控制了城市建设强度。

STEP.2 构建生态绿网
保留内部山体，利用与周边山体呼应的山体绿色通廊将两个集水流域的水体串联，构建山水交错的"十字型"生态网络，形成覆盖各年龄段和社会阶层，促进交流融通的人气公共活力网络。

STEP.3 强化绿网关系
在生态绿网的基础上，打造休闲慢行网络，强化绿网关系。打造以水为主体的纵向步道与以山体斑块为主体的横向步道，纵横步道在城市最为核心的部位进行结合。

STEP.4 激活公共中心
结合生态网络，规划城市活力公共中心，集中塑造滨水地区的商业、文化、娱乐、公共职能，强调滨水的公共属性，通过汇聚多元功能为栋溪河带来 24 小时的持久吸引力。

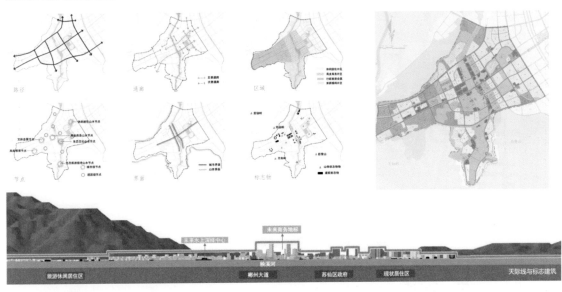

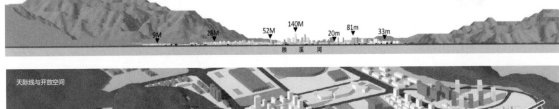

137

总平面图（低碳生态产业园城市设计）

6

设计题目： 低碳生态产业园城市设计
指导老师： 焦胜　沈瑶
学　　生： 云华杰

Design topic: Urban Design of Low Carbon Ecological Industrial Park

Instructors: Jiao Sheng,Shen Yao

Student: Yun Huajie

● 设计说明

近年来，由于城市对山水林田湖的保护不足，开发强度高、硬质铺装多等原因，下垫面过度硬化，改变了城市原有自然生态本底和水文特征，破坏了自然的"海绵体"，导致"逢雨必涝、雨后即旱"，城市看海成为威胁人民生命财产安全、阻碍城市发展的主要危害之一，同时也带来了水环境污染、水资源紧缺、水安全缺乏保障、水文化消失等一系列问题。

设计通过对海绵城市的研究，进行以生态建设为基底的生态低碳规划设计，合理利用自然地形、自然水系，采用低影响的开发模式，优先选用生态型措施，尽可能多预留城市绿地空间，增加可渗透地面，蓄积雨水宜就地回用，保留原有场地雨水径流和生态基底，降低建设成本，改变传统设计方式。设计过程中除了对功能、交通等因素的考虑，同时在分析阶段采取量化分析的技术手段，对自然地形和生态要素进行综合考虑，保证设计的科学性。

Design notes

In recent years, due to the insufficient protection of mountains, rivers, forests, farmlands and lakes, and the high development intensity and many hard pavements, the underlying surface has been excessively hardened, which has changed the original natural ecological background and hydrological characteristics of the city, damaged the natural "sponge", resulting in "waterlogging in case of rain and drought after rain". "Looking at the sea in the city" has become a threat to the safety of people's lives and property, And it is also one of the main hazards hindering urban development. At the same time, it also brings a series of problems, such as water environmental pollution, shortage of water resources, lack of guarantee of water security, disappearance of water culture and so on.

Through the study of Sponge City, the design carries out ecological low-carbon planning and design based on ecological construction, makes rational use of natural terrain and natural water system, adopts low impact development mode, gives priority to ecological measures, reserves urban green space as much as possible, increases permeable ground, and the accumulated rainwater should be reused on site, so as to retain the rainwater runoff and ecological base of the original site, reduce the construction cost and change the traditional design method. In the design process, in addition to considering the factors such as function and traffic, the technical means of quantitative analysis shall be adopted in the analysis stage to comprehensively consider the natural terrain and ecological elements to ensure the scientificity of the design.

城市公园、景色优美、西侧湖、东侧河、北侧海创园展示中心地标建筑，入口活动广场，树林，步道，游乐场，骑行，公共活动形式多样　　　　**公园效果透视**

● 基地分析 Base analysis

项目位于长沙市高新区雷锋片区，北临金州大道，南面枫林西路，西接望雷大道，东接雷高路，东北部与雷锋河相接，总用地面积 1.9km² 左右，标高 38.1~94.2m，现状为农业用地，以丘陵、池塘与农田为主。本设计在地块范围内选择 0.5~1.0km² 进行城市设计。

The project is located in Leifeng area of Changsha high tech Zone, adjacent to Jinzhou Avenue in the north, Fenglin West Road in the south, Wanglei Avenue in the west, Leigao road in the East, and Leifeng River in the northeast. The total land area is about 1.9 square kilometers, with an elevation of 38.1~94.2. It is currently agricultural land, mainly hills, ponds and farmland. In this design, 0.5~1.0 square kilometers are selected for urban design within the plot.

设计过程中除了对功能、交通等因素的考虑，同时在分析阶段采取量化分析的技术手段，对自然地形和生态要素进行综合考虑。通过生态规划合理利用场地自有雨水资源，实现可持续雨污循环，建设生态产业园。

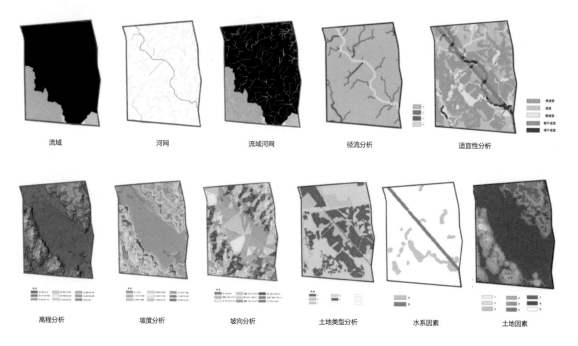

流域　　　　　　河网　　　　　　流域河网　　　　　径流分析　　　　　适宜性分析

高程分析　　　坡度分析　　　坡向分析　　　土地类型分析　　　水系因素　　　土地因素

● 道路、水系与生态　Roads, water systems and ecology

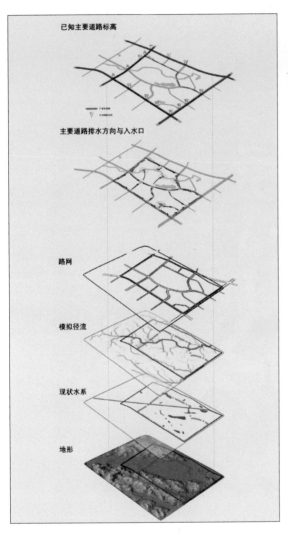

路网生成

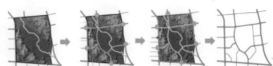

本次路网规划是在现状道路的基础上，对科技城原本规划道路进行修改，改变其传统道路规划成果，适应地形与生态景观需求，形成方格网与自由式结合的通达的道路网系统。

水系修复

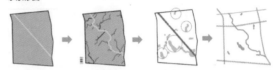

保留场地中部分原有的池塘，作为蓄水或景观空间，结合模拟径流和现状径流，适当恢复并拓宽截弯取直前的河道。在水系的基础上，结合路网的布置添加绿地，形成最终的生态格局。

生态格局

路网生成

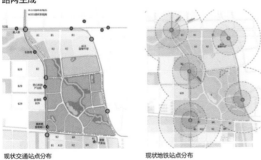

现状交通站点分布　　　　　　　现状地铁站点分布

现状交通站点分布不均匀，场地内部无公共交通站点分布。因此，沿着公交线路，尤其是在站点周边土地高强度开发，公共使用优先。公共功能主要集中在核心腹地，直接范围形成复合开发间地。现状地块范围内只有三条公交线路通过，其中雷高路及地块内部没有可达的公交站点，地块很大范围未在站点服务半径范围之内。

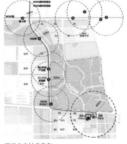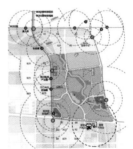

现状公交站点分布　　　　　　　综合分析

通过上述分析，确定规划策略：（1）在500米半径的步行影响范围内，雷锋河以东大部分地块未在服务范围内，建议增设站点，实现地块全覆盖。（2）站点周围用地布局宜采用开发高密度住宅、商业、办公用地，同时开发服务业、娱乐、体育等公共设施的混合利用模式。

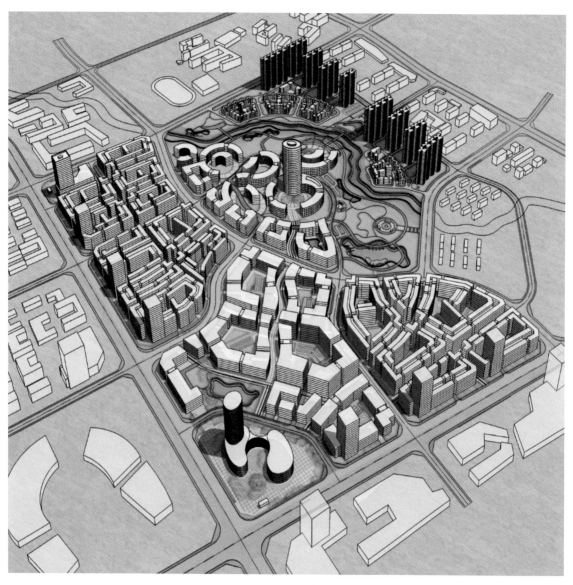

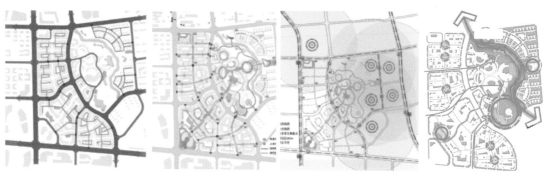

● 指标控制 Index control

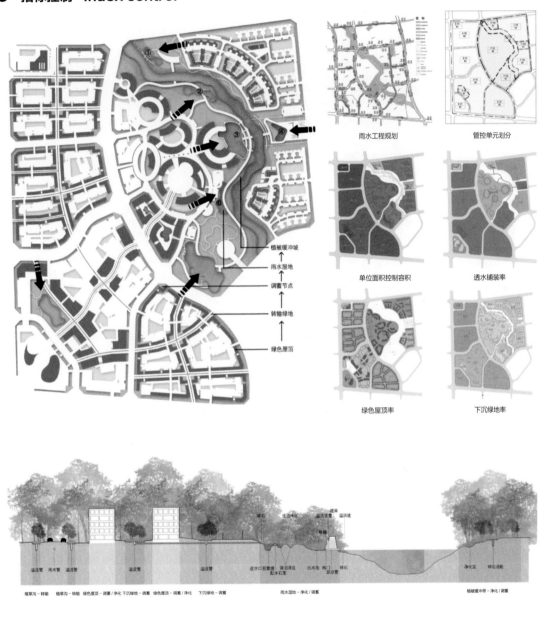

植被缓冲坡
雨水湿地
调蓄节点
转输绿地
绿色屋顶

雨水工程规划

管控单元划分

单位面积控制容积

透水铺装率

绿色屋顶率

下沉绿地率

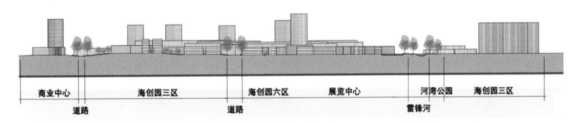

植草沟－转输　植草沟－转输　绿色屋顶－调蓄／净化 下沉绿地－调蓄　绿色屋顶－调蓄／净化　下沉绿地－调蓄　　　　　雨水湿地－净化／调蓄　　　　植被缓冲带－净化／调蓄

溢流管　雨水管　溢流管　　　溢流管　　　溢流管　　　　　　碎石　浅溏湿区　溢流竖管　溢洪道　　　　　　　　　　净化区　碎石消能
　　　　　　　　　　　　　　　　　　　　　　　　透水口前置塘　深沼泽区　出水池　闸门　　　　　　　　　　　　　　碎石
　　　　　　　　　　　　　　　　　　　　　　　　　配水石笼　　　　　　放空管

商业中心	海创园三区	海创园六区	展览中心	河湾公园	海创园三区
道路		道路		雷锋河	

专题三：村镇规划设计
Topic 3: Village Planning and Design

总平面图（乌龙堤村人居环境规划）

设计题目： 资阳区乌龙堤村美丽乡村人居环境规划

指导老师： 姜敏　周恺

学　　生： 何磊

7

Design topic: Living Environment Planning of "Beautiful Village" in Wulongdi Village, Ziyang District

Instructors: Jiang Min,Zhou Kai

Student: He Lei

● 设计说明

乌龙堤村由注南湖村、下资口村、长泊湖村组成，因为1999年洞庭湖流域洪水，三个村庄合并成为一个村，并在下资口村村口堆高地设立集中安置点。

着眼于乌龙堤村在当今时代背景下的前景，寻求符合时代发展要求和本村村民发展需求的道路，从实际出发，结合国家针对农村问题提出的各项政策，切实落实各项措施，做好农村产业布局，加强乡村基础设施建设和环境整治，切实改善农村的生产生活环境，真正让乡村"看得见山，望得见水，记得住乡愁"，真正推动乌龙堤村更好、更科学地发展。

Design notes

Wulongdi Village is composed of Zhunanhu Village, Xiazikou Village, and Changbohu Village. Because of the flood in the Dongting Lake basin in 1999, the three villages merged into one village, and a centralized resettlement site was set up at the entrance to Xiazikou Village.

Focusing on the prospect of Wulongdi village in the current era, seeking a path that meets the development requirements of the times and the development needs of the villagers in the village, starting from reality and in combination with various policies put forward by the state for rural problems, we need to earnestly implement various measures, do a good job in rural industrial layout, strengthen rural infrastructure construction and environmental improvement, and effectively improve the production and living environment in rural areas, so as to achieve the listen of "see the mountains and waters in the village and keep mostalgia in mind",and truly promote the better and more scientific development of Wulongdi village.

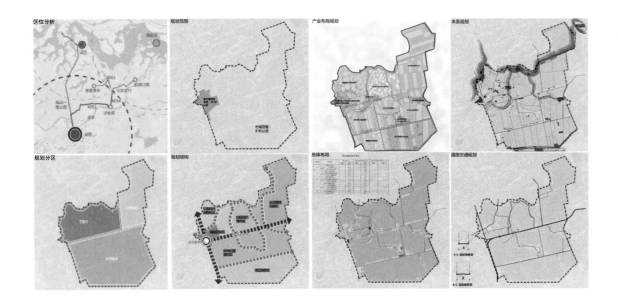

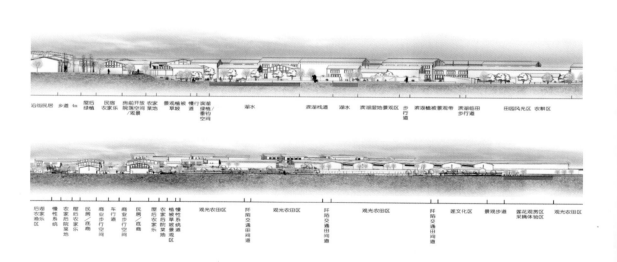

沿街民居　乡道 4m　屋后绿植　民宿农家乐　房前开敞院落空间/观景　农家菜地　景观植被草坡　慢行道　滨湖绿植/垂钓空间　湖水　滨湖栈道　湖水　滨湖湿地景观区　步行道　滨湖植被景观带　滨湖临田步行道　田园风光区　农耕区

湖后农渔乐区　慢性系统　屋后农家院菜地　民居/底商　车行道　商业步行空间　民居/底商　屋后农家院菜地　慢性步行空间　植被草系统坡景统观区　观光农田区　阡陌交通田间道　观光农田区　阡陌交通田间道　观光农田区　阡陌交通田间道　莲文化区　景观步道　莲花观赏区采摘体验区　观光农田区

● 规划设计分析　Planning and design analysis

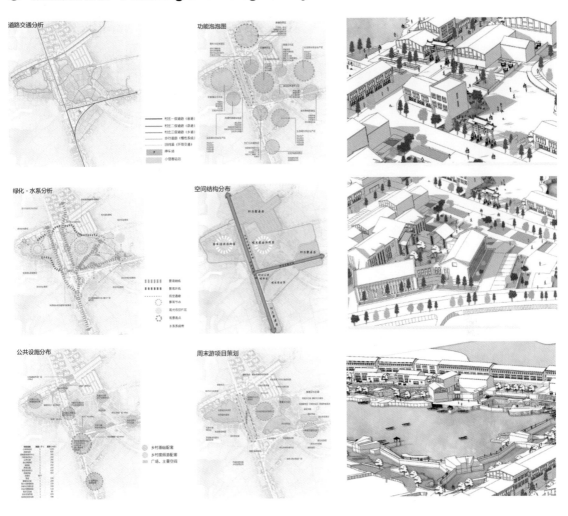

道路交通分析

功能泡泡图

绿化 - 水系分析

空间结构分布

公共设施分布

周末游项目策划

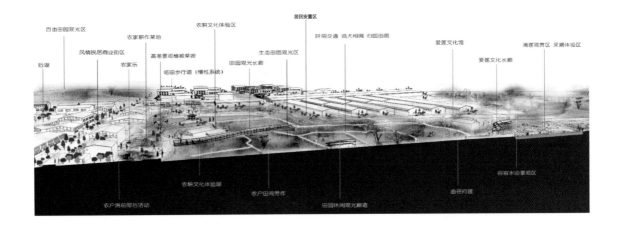

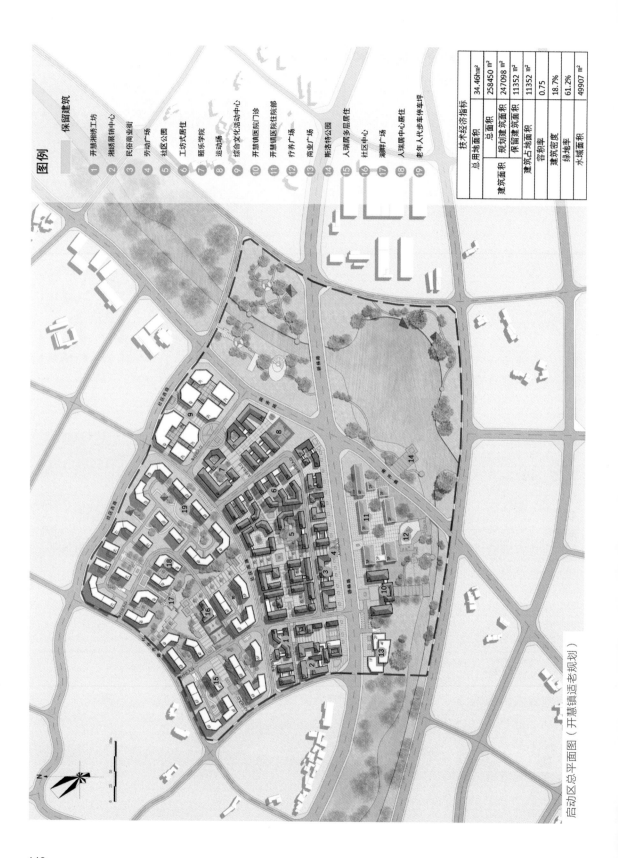

技术经济指标		
总用地面积		34.46hm²
建筑面积	总面积	258450 m²
	规划建筑面积	247098 m²
	保留建筑面积	11352 m²
保留建筑面积		11352 m²
建筑占地面积		
容积率		0.75
建筑密度		18.7%
绿地率		61.2%
水域面积		49907 m²

图例

保留建筑

① 开慧湘绣工坊
② 湘绣展销中心
③ 民俗商业街
④ 劳动广场
⑤ 社区公园
⑥ 工坊式居住
⑦ 颐乐学院
⑧ 运动场
⑨ 综合文化活动中心
⑩ 开慧镇医院门诊
⑪ 开慧镇医院住院部
⑫ 疗养广场
⑬ 商业广场
⑭ 斯洛特公园
⑮ 人瑞居多层居住
⑯ 社区中心
⑰ 湖畔广场
⑱ 人瑞居中心居住
⑲ 老年人代步车停车坪

启动区总平面图（开慧镇适老规划）

148

8

设计题目： 开慧镇适老规划与设计研究
指导老师： 邱灿红　丁国胜
学　　生： 陆筱恬

Design topic: Planning and Design of Suitable Space for the Elderly in Kaihui Town

Instructors: Qiu Canhong,Ding Guosheng

Student: Lu Xiaotian

● **设计说明**

长沙市养老设施数量不足，分布不均，老城区养老压力显著。（1）长沙老城区养老设施建设类型较单一；（2）长沙老城区养老设施配备床位数远少于需求，老龄化带来养老相关需求的攀升，城市养老设施供不应求，尤其是老城区老年人分布集中，但建设用地不足，养老压力显著；（3）养老设施分布情况和老年人分布不匹配，老城区养老设施压力显著。

Design notes

The number of elderly care facilities in Changsha is insufficient and unevenly distributed, so the elderly care pressure in the old urban area is significant. 1. The construction type of elderly care facilities in the old urban area of Changsha is relatively single. 2. The number of beds equipped with elderly care facilities in the old urban area of Changsha is far less than the demand. The aging has brought about an increase in the demand for elderly care. The supply of urban elderly care facilities is in short supply, especially the elderly in the old urban area are concentrated, but the construction land is insufficient, and the elderly care pressure is significant. 3. The distribution of elderly care facilities does not match the distribution of the elderly, and the pressure on elderly care facilities in the old urban area is significant.

● **研究框架　Research framework**

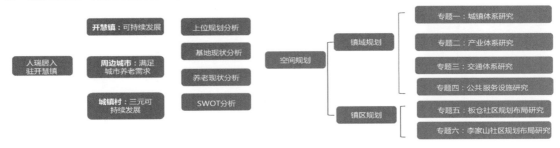

● 区位分析　Location analysis

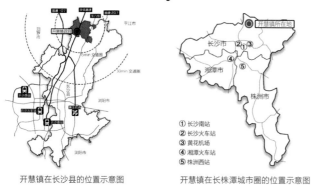

① 长沙南站
② 长沙火车站
③ 黄花机场
④ 湘潭火车站
⑤ 株洲西站

开慧镇在长沙县的位置示意图　　　　开慧镇在长株潭城市圈的位置示意图　　　　开慧镇在长沙市的位置示意图

● 痛点剖析　Pain point analysis

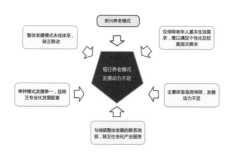
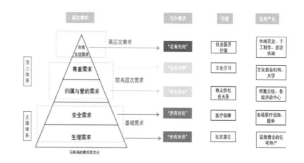

● 现状分析　Current situation analysis

镇域人口现状分布　　　　镇域人口规划分布　　　　镇域村庄适老化等级评价　　　　镇域村镇体系规划

● 规划结构　Planning structure

镇域适老产业布局规划　　　　镇域适老产业重点项目规划　　　　镇域适老重点项目游线规划　　　　镇域土地利用规划

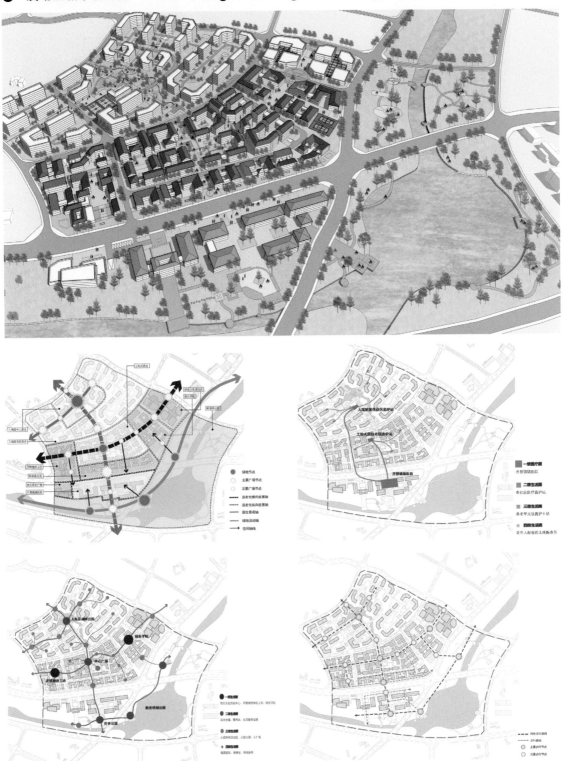

鸟瞰图（双牌县茶林镇区城市设计）

9

设计题目： 双牌县茶林镇镇区城市设计
指导老师： 焦胜
学　　生： 陈雅湄

Design topic: Urban Design of Chalin Town, Shuangpai County

Instructor: Jiao Sheng

Student: Chen Yamei

● 设计说明

本设计基地茶林镇位于双牌县东北山区，东接阳明山，南连麻江镇。为了响应国家特色小镇政策的推进以及阳明山景区和永州乡村振兴区的发展，本设计将运用城市规划的技术手段，系统分析茶林镇区的基本特征，从区域层面对茶林特色小镇进行全面的分析，在透彻分析地块土地利用现状、结合相关规划以及特色小镇等发展特点的基础上，提出了茶林特色镇区的发展目标及产业发展策略，同时合理组织镇区土地使用空间布局，完善基础设施布置，并对小镇的城市空间形态进行设计，其中包括镇区整体城市设计及邓家大院核心景区的详细设计。

Design notes

Chalin Town, the design base, is located in the northeast mountainous area of Shuangpai County, connecting Yangming Mountain in the East and Majiang town in the south. In response to the promotion of the national characteristic town policy and the development of Yangming Mountain scenic spot and Yongzhou Rural Revitalization zone, the design will systematically analyze the basic characteristics of Chalin town by using the technical means of urban planning, comprehensively analyze the Chalin characteristic town from the regional level, thoroughly analyze the land use status of the plot, combined with relevant planning based on the development characteristics of Chalin characteristic town, this paper puts forward the development objectives and industrial development strategy of Chalin characteristic town, reasonably organizes the land use space layout of the town and improves the infrastructure layout, and designs the urban space form of the town. It includes the overall urban design of the town and the detailed design of the core scenic spot of the Deng's courtyard.

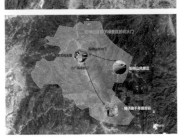

区位分析图

山水格局分析图

● 基地现状 Current state of the base

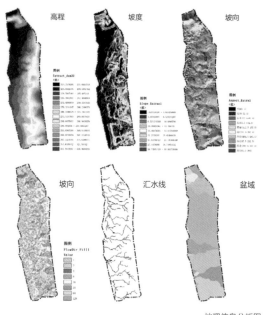

高程　　坡度　　坡向

坡向　　汇水线　　盆域

地理信息分析图

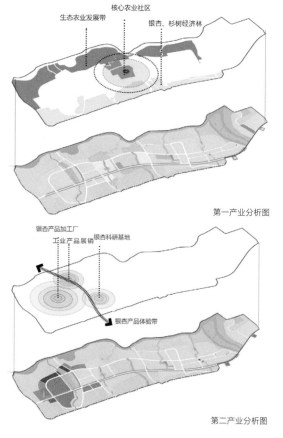

茶林

建设用地
耕地
水系
林地
道路用地

土地利用分析图

茶林中学　变电站　　茶林乡政府
　　　　　　　　　　茶林乡市场
茶林幼儿园　交警大队　汽车站　茶林乡医院
　　　　　　　　　　　　　工商管理部

公共服务设施分析图

核心农业社区
生态农业发展带　　　　银杏、杉树经济林

第一产业分析图

G55 二广高速
永连最美公路

道路交通分析图

银杏产品加工厂
工业产品展销　银杏科研基地

银杏产品体验带

第二产业分析图

154

控制性详细规划图 Regulatory detailed plan

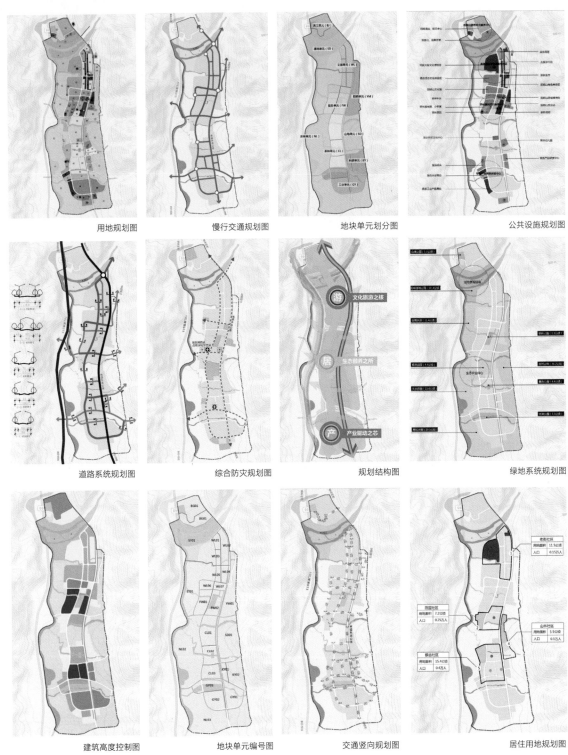

用地规划图　　　　　　　慢行交通规划图　　　　　　地块单元划分图　　　　　　公共设施规划图

道路系统规划图　　　　　　综合防灾规划图　　　　　　规划结构图　　　　　　　　绿地系统规划图

建筑高度控制图　　　　　　地块单元编号图　　　　　　交通竖向规划图　　　　　　居住用地规划图

● 规划策略 Planning strategy

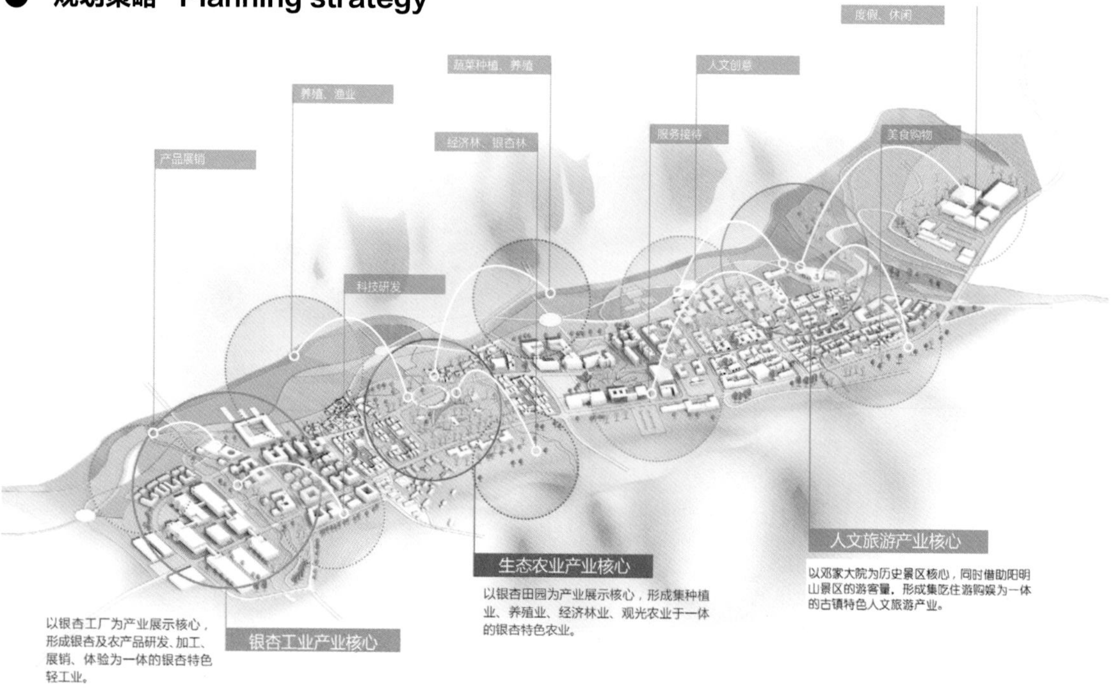
度假、休闲
蔬菜种植、养殖
人文创意
养殖、渔业
服务接待
经济林、银杏林
美食购物
产品展销
科技研发
人文旅游产业核心
以邓家大院为历史景区核心，同时借助阳明山景区的游客量，形成集吃住游购娱为一体的古镇特色人文旅游产业。
生态农业产业核心
以银杏田园为产业展示核心，形成集种植业、养殖业、经济林业、观光农业于一体的银杏特色农业。
以银杏工厂为产业展示核心，形成银杏及农产品研发、加工、展销、体验为一体的银杏特色轻工业。
银杏工业产业核心

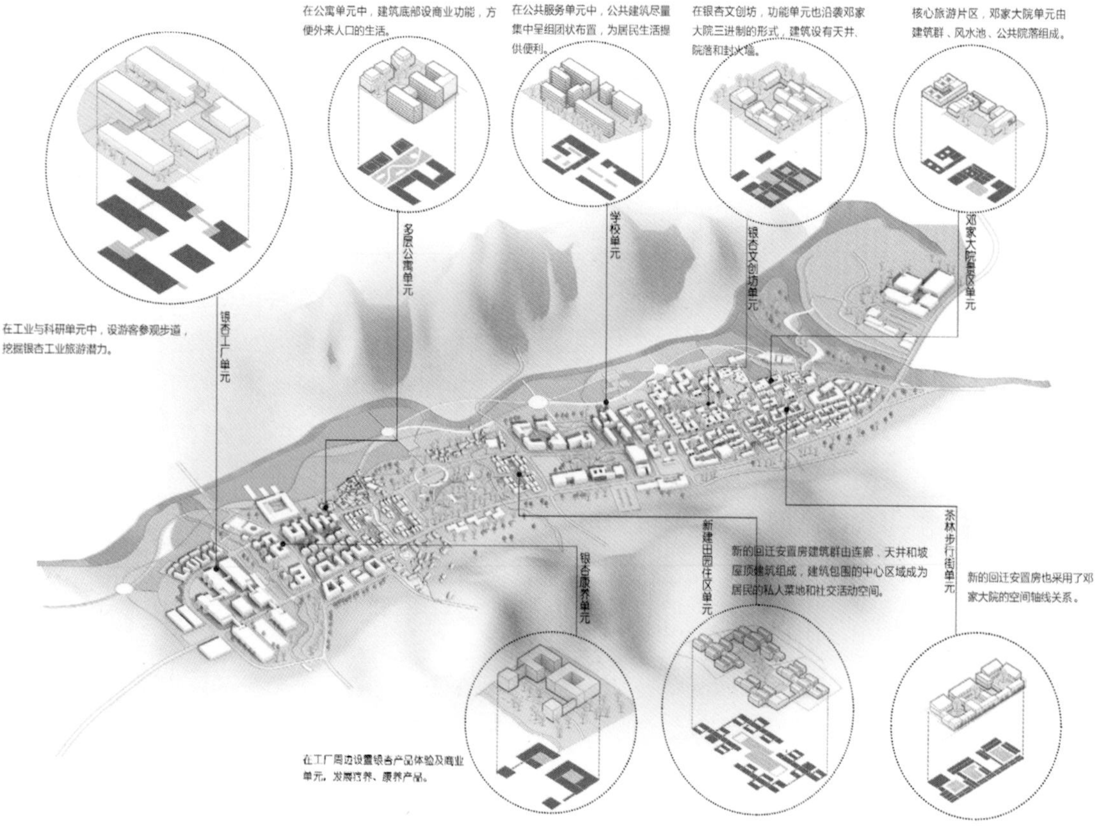
在公寓单元中，建筑底部设商业功能，方便外来人口的生活。
在公共服务单元中，公共建筑尽量集中呈组团状布置，为居民生活提供便利。
在银杏文创坊，功能单元也沿袭邓家大院三进制的形式，建筑设有天井、院落和封火墙。
核心旅游片区，邓家大院单元由建筑群、风水池、公共院落组成。
多层公寓单元
学校单元
银杏文创坊单元
邓家大院景区单元
在工业与科研单元中，设游客参观步道，挖掘银杏工业旅游潜力。
银杏工厂单元
新建田园住区单元
新的回迁安置房建筑群由连廊、天井和坡屋顶建筑组成，建筑包围的中心区域成为居民的私人菜地和社交活动空间。
茶林步行街单元
新的回迁安置房也采用了邓家大院的空间轴线关系。
银杏康养单元
在工厂周边设置银杏产品体验及商业单元，发展疗养、康养产品。

总平面图 General layout

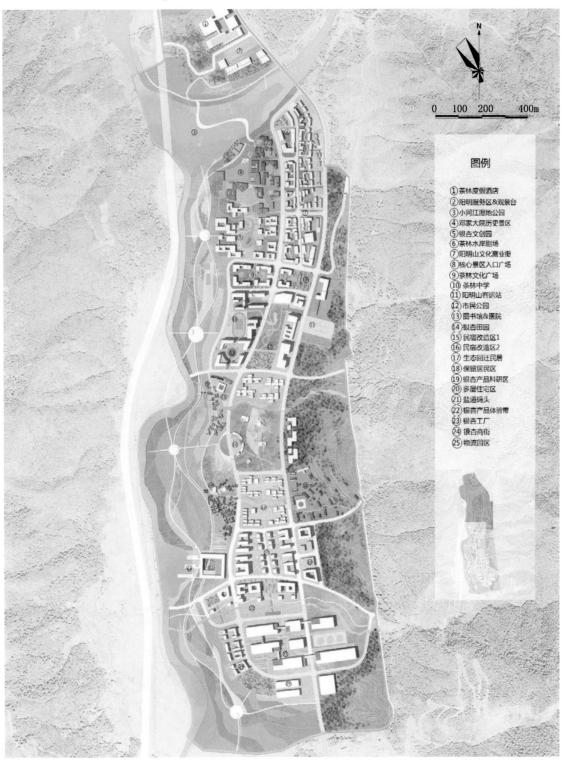

N

0 100 200   400m

图例

① 茶林度假酒店
② 阳明服务区&观景台
③ 小河江湿地公园
④ 邓家大院历史景区
⑤ 银杏文创园
⑥ 茶林水岸剧场
⑦ 阳明山文化商业街
⑧ 核心景区入口广场
⑨ 茶林文化广场
⑩ 茶林中学
⑪ 阳明山客运站
⑫ 市民公园
⑬ 图书馆&医院
⑭ 银杏田园
⑮ 民宿改造区1
⑯ 民宿改造区2
⑰ 生态回迁民居
⑱ 保留居民区
⑲ 银杏产品科研区
⑳ 多层住宅区
㉑ 盐道码头
㉒ 银杏产品体验带
㉓ 银杏工厂
㉔ 银杏商街
㉕ 物流园区

核心景区设计 Core scenic spot design

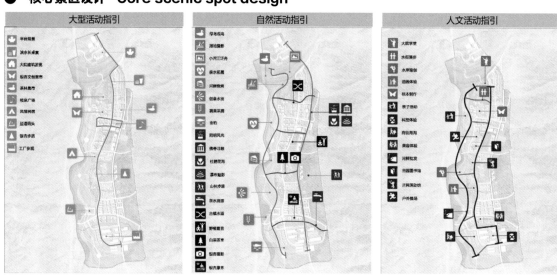

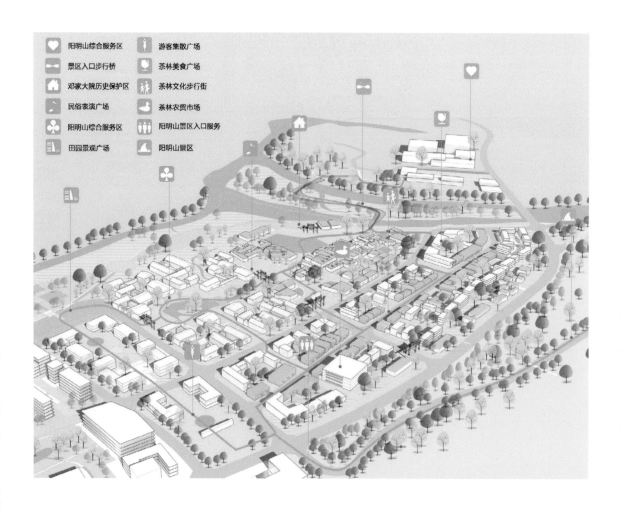

158

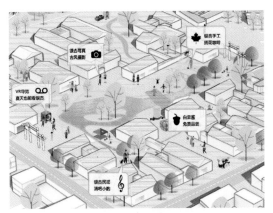

银杏休闲文创坊

书吧、茶室、手工、
清吧、摄影、文创、
戏曲、音乐、展览

人群 / 推荐指数

中老年人 ★★★☆☆		一日游 ★★★☆☆
青年人 ★★★★★		深度游 ★★★★☆
儿童 ★★★★☆		跟团游 ★★★☆☆

文创坊由四个庭院建筑群组成，是集民宿、清吧、茶室、手作、艺术活动为一体的休闲街区。

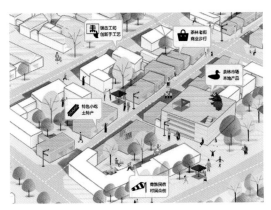

茶林步行街

特产、美食街、小吃、
白果、家禽、糕点、
咖啡、老建筑

人群 / 推荐指数

中老年人 ★★★☆☆		一日游 ★★★★☆
青年人 ★★★☆☆		深度游 ★★☆☆☆
儿童 ★★★☆☆		跟团游 ★★★☆☆

保留原有茶林老街，改造升级店铺立面，打造具有历史人文气息的步行街。

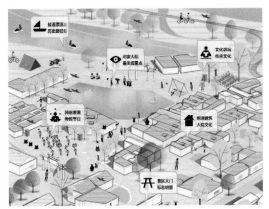

邓家大院历史保护区

建筑、学堂、民俗、
表演、纪念品

人群 / 推荐指数

中老年人 ★★★★★		一日游 ★★★★★
青年人 ★★★☆☆		深度游 ★★★★☆
儿童 ★★★★★		跟团游 ★★★★☆

来自阳明山综合服务区的游客可以通过河滩上的步行桥直接抵达邓家大牌坊、风水池、建筑群，三点一线成为最佳摄影点，是历史景区的核心空间。

广场视觉通廊

旅游讲解、物品存放、
设备租赁、户外拓展、
团队建设、亲子活动

人群 / 推荐指数

中老年人 ★★★☆☆		一日游 ★★★★☆
青年人 ★★★★☆		深度游 ★★★☆☆
儿童 ★★★★☆		跟团游 ★★★★★

阳明山入口西侧由游客集散广场、旅游办公区、佛教文化展览馆组成的轴线建筑群，多个广场和水田通过阶梯相互贯通。

专题四：城市特色片区更新规划与设计
Topic 4: Renewal Planning and Design of Urban Characteristic Areas

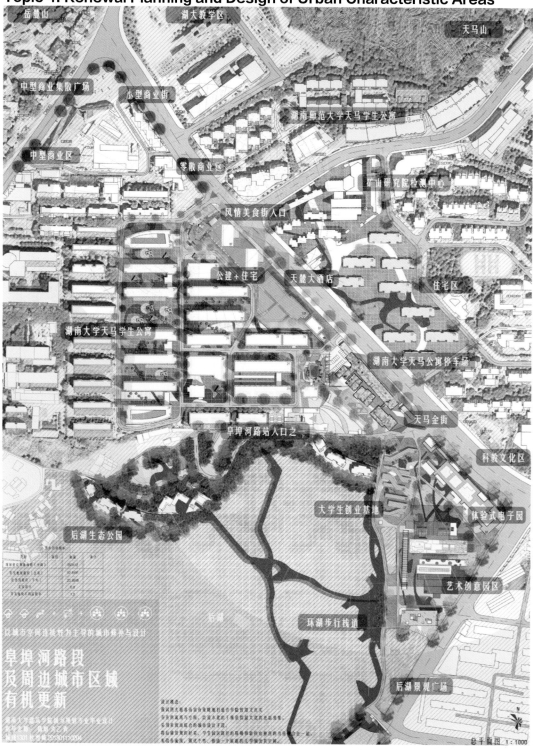

设计题目： 岳麓区阜埠河路段及周边城市区域有机更新
指导老师： 陈煊　许乙青
学　　生： 杜雪峰

10

Design topic: Organic Renewal of Fubu River Section and the Surrounding Areas

Instructors: Chen Xuan,Xu Yiqing

Student: Du Xuefeng

● 设计说明

用规划的思路去发现问题，用设计的手法去解决问题。就好比修理一个巨大的精密仪器，你需要从宏观上知道它运行的原理，知道它运作的规律，而当你真正去修补它时，需要我们做的可能是改变某一个齿轮的形状、某一个螺母的形态这样细节的操作。

Design notes

Use planning ideas to find problems and design methods to solve problems. Just like repairing a huge and precise instrument, you need to know its operation principle and operation law from the macro level. When you really repair it, what we need to do may be something specific, like change the shape of a gear or the shape of a nut.

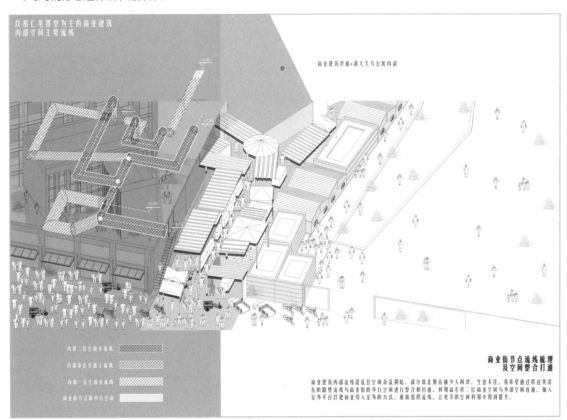

●效果图 Effect picture

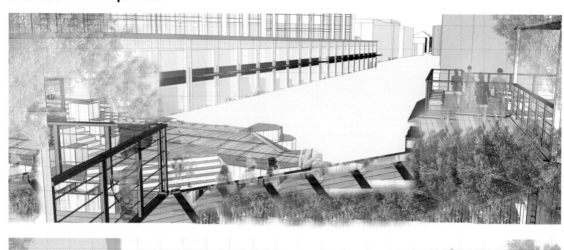

● 鸟瞰图 Aerial view

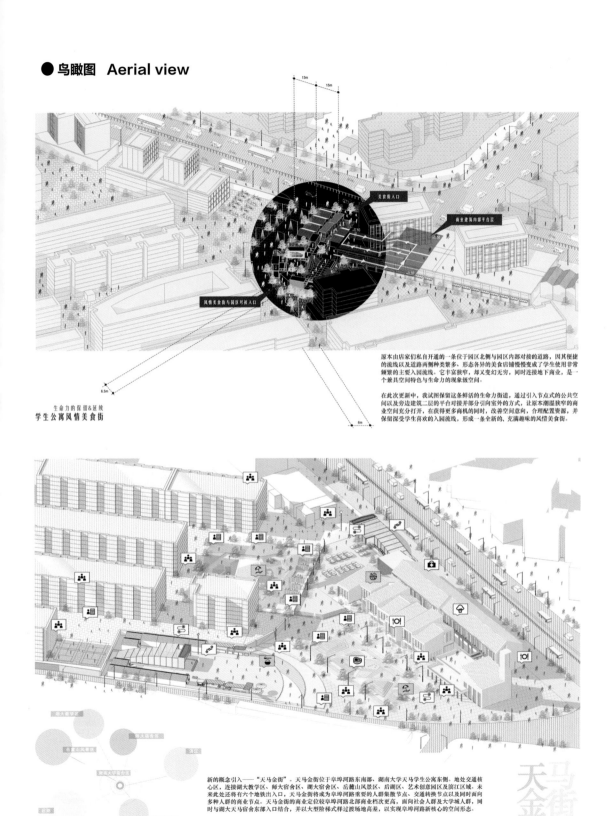

13m   15m

美食街入口

商业建筑内部平台层

风情美食街与园区另一入口

6.5m

6m

生命力的保留&延续
学生公寓风情美食街

原本由店家们私自开通的一条位于园区北侧与园区内部对接的道路，因其便捷的流线以及道路两侧类似多，形态各异的美食店铺慢慢变成了学生使用非常频繁的主要入园流线。它丰富狭窄，却又变幻无穷，同时连接地下商业，是一个兼具空间特色与生命力的现象级空间。

在此次更新中，我试图保留这条鲜活的生命力街道，通过引入节点式的公共空间以及旁边建筑二层的平台对接并部分引向室外的方式，让原本潮湿狭窄的商业空间充分打开，在获得更多商机的同时，改善空间意向，合理配置资源，并保留深受学生喜欢的入园流线，形成一条全新的、充满趣味的风情美食街。

天马金街

湖大教学区
师大宿舍区
岳麓山风景区
滨北
湖大宿舍东区
艺术创意园区

新的概念引入——"天马金街"。天马金街位于阜埠河路东南部、湖南大学天马学生公寓东侧。地处交通核心区，连接湖大教学区、师大宿舍区、湖大宿舍区、岳麓山风景区、后湖区、艺术创意园区及滨江区域。未来此处还将有六个地铁出入口，天马金街将成为阜埠河路重要的人群集散节点、交通转换节点以及同时面向多种人群的商业节点。天马金街的商业定位较阜埠河路北部商业档次更高，面向社会人群及大学城人群，同时与湖大天马宿舍东部入口结合，并以大型阶梯式样过渡场地高差，以实现阜埠河路新核心的空间形态。

阜埠河路一体式新街道转换空间

163

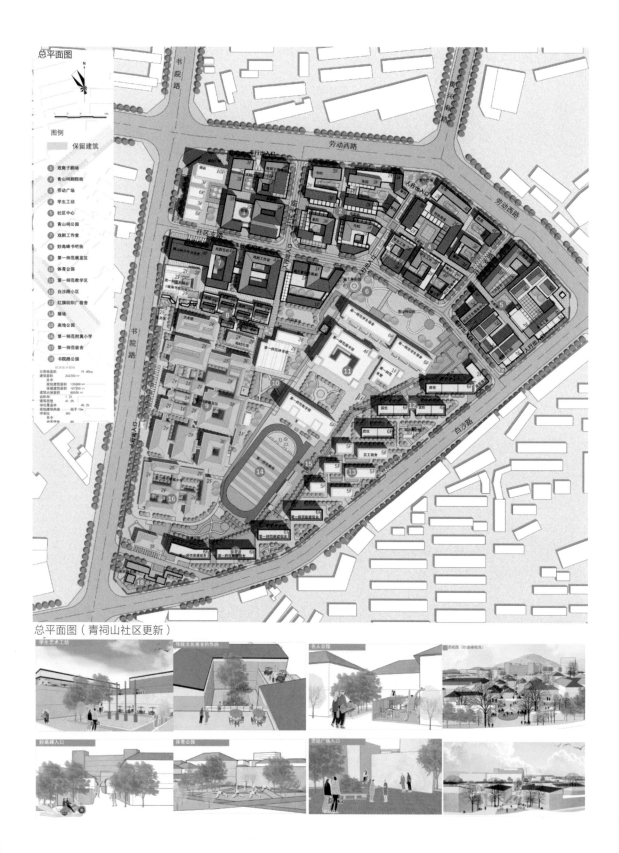

总平面图（青祠山社区更新）

11

设计题目： 长沙青祠山社区更新设计
指导老师： 丁国胜　邱灿红
学　　生： 周晓穗

Design topic: Renewal Design of Qingcishan Community in Changsha

Instructors: Ding Guosheng, Qiu Canhong

Student: Zhou Xiaosui

● 设计说明

本设计着重打造社区内部交通系统、步行系统、开放空间系统，注重社区景观设计。为创造有生气的街道界面和优美的街景停车，机动车出入口等被隐藏在主体建筑之后。步行系统中规划社区特色步行路线，承载重要步行节点，突出青山祠社区的文化特色。开放空间被分为四类加以充分利用，营造浓厚的社区活动氛围。在社区景观设计上，则通过打造视觉焦点与文化节点，着重建立起两者在空间上的联系。

Design notes

This Design focuses on building the Community Internal Transportation System, Pedestrian System and Open Space System, and pays attention to the Community Landscape Design. In order to create a lively street interface and beautiful street view parking, Motor Vehicle Entrances and Exits are hidden behind the main building. In the Pedestrian System, the Community Characteristic Pedestrian Route is planned to carry important Pedestrian Nodes and highlight the Cultural Characteristics of Qingshanci Community. Open Space is divided into four categories to make full use of and create a strong atmosphere of Community Activities. In the Community Landscape Design, we focus on establishing the Spatial Connection between the two by creating Visual Focus and Cultural Nodes.

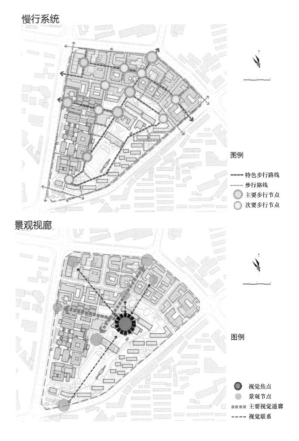

交通系统

图例
━━ 主要主干道
── 主干道
── 次干道
── 社区支路
── 步行道
● 地面停车
Ⓟ 地下停车

慢行系统

图例
┄┄ 特色步行路线
── 步行路线
◉ 主要步行节点
○ 次要步行节点

开放空间系统

图例
休闲公园
校园绿地
邻里绿地
活动小广场
体育公园

景观视廊

图例
◉ 视觉焦点
● 景观节点
▪▪▪ 主要视觉通廊
┄┄ 视觉联系

● 测度指标建立　Establishment of measurement index

根据现有建成环境测度的研究情况并且考虑到本次测度
的可行性，我们对 33 个社区建立了以下测度指标体系。
体系下共有 5 个主因子，分别为社会环境、生态环境、
交通环境、服务设施、建设环境。社会环境主要考虑人
口密度、外来人口占比和社区封闭度；生态环境主要考虑
卫生环境因素、绿化覆盖率和地形起伏度；交通环境考虑
公交可达性和道路通达度；服务设施考虑商业服务设施、
教育设施、文化体育设施和医疗设施；建筑环境包括建筑
高度、建筑年代、建筑密度和开发强度。

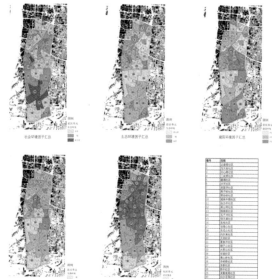

● 社区特征分类　Classification of community characteristics

通过 SPSS 聚类分析五大主因子，归纳出六类社区类型，
分别为：老旧稳定社区、发展融合社区、衰败退化社区、
封闭冲突社区、丘陵特色潜力社区、开放和谐社区。其中，
老旧稳定社区有 11 个；发展融合社区有 4 个；衰败退化
社区有 7 个；封闭冲突社区有 4 个；丘陵特色潜力社区有
5 个；开放和谐社区有 1 个，详细见下表。

社区类型	包含社区	空间类型	空间特征	更新策略
第一类社区空间	工农村社区　游湖社区　燕子村社区　绿谷广场社区　锦绣街社区　石子冲社区　天剑社区　直家冲社区　狮子山社区　长城社区　吴家岭社区	老旧稳定社区	社区服务各项及总体评分较稳定，老旧小区为主，但内部交通通达性差	步行交通系统的构建
第二类社区空间	火星球桥社区　鸭子铺社区　辣椒子桥社区　化龙池社区	发展融合社区	21世纪初建成的发展成熟小区为主。社区环境开放，商业评分偏高，社区服务分异性低	结合社区特点进行更新改造
第三类社区空间	太葵春社区　涂新社区　天心阁社区　向东南社区　大巡山社区　黄土岭社区　新开家园社区　南大桥社区	衰败退化社区	社区人口密度低，社区服务设施指标偏低	综合改造，社区整治，增加服务设施
第四类社区空间	白沙井社区　宝塔山社区　古柏路社区　新下社区	封闭冲突社区	封闭度高，社区服务分异性大	为不同人群提供多样活动的机会，促进不同人群交往融合
第五类社区空间	丘陵社区　沙河社区　古树岭社区　东电社区　东屋山社区	丘陵特色潜力社区	绿化环境差，丘陵特色明显，人口密度高，老旧社区为主，交通便利	环境整治；同时利用本地居民偏多的优势，发掘老长沙文化；利用丘陵地形特色，凸显长沙丘陵特色
第六类社区空间	白沙花园社区	开放和谐社区	社区规模小，管理便利，开放社区，各项指标偏优	社区模式推广，共享发展正外部性

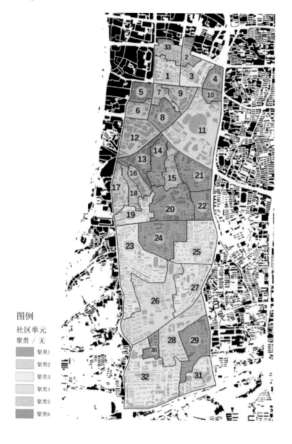

图例
社区单元
聚类 / 无
聚类1
聚类2
聚类3
聚类4
聚类5
聚类6

166

● 整体更新评价与意向　Evaluation and intention of overall renewal

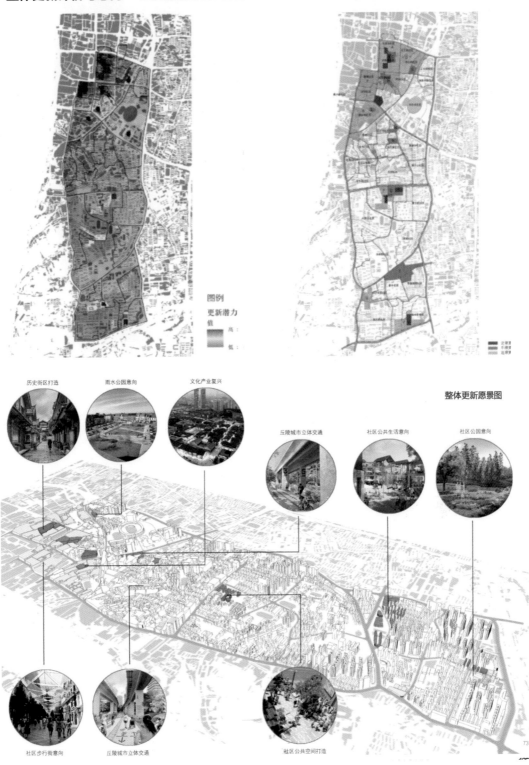

图例
更新潜力值

历史街区打造　　湖水公园意向　　文化产业复兴

整体更新愿景图

丘陵城市立体交通　社区公共生活意向　社区公园意向

社区步行街意向　　丘陵城市立体交通　　社区公共空间打造

73

167

专题五：“旧城栖居”联合毕业设计
Topic 5: Historical Trail Update Design

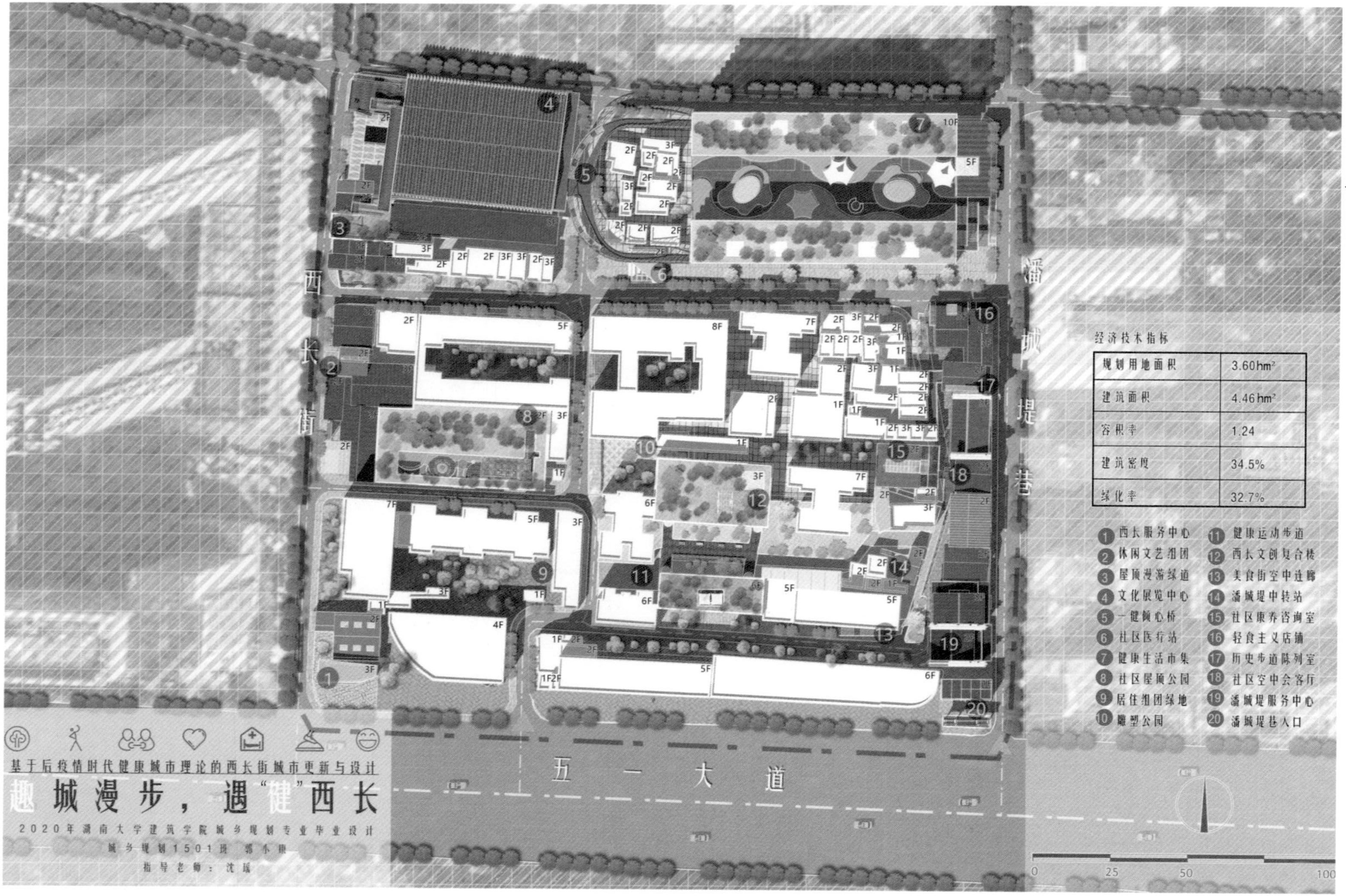
基于后疫情时代健康城市理论的西长街城市更新与设计
趣城漫步，遇“健”西长
2020年湖南大学建筑学院城乡规划专业毕业设计
城乡规划1501班 郭小康
指导老师：沈瑶
经济技术指标
规划用地面积 3.60hm²
建筑面积 4.46hm²
容积率 1.24
建筑密度 34.5%
绿化率 32.7%
1 西长服务中心
2 休闲文艺组团
3 屋顶漫游绿道
4 文化展览中心
5 一健倾心桥
6 社区医疗站
7 健康生活市集
8 社区屋顶公园
9 居住组团绿地
10 雕塑公园
11 健康运动步道
12 西长文创复合楼
13 美食街空中连廊
14 潘城堤中转站
15 社区康养咨询室
16 轻食主义店铺
17 历史步道陈列室
18 社区空中会客厅
19 潘城堤服务中心
20 潘城堤巷入口
西长街
潘城堤巷
五一大道
0 25 50 100

设计题目： 趣城漫步，遇"健"西长——四校联合毕业设计
指导老师： 沈瑶
学　　生： 郭小康

Design topic: Walking in the Interesting City, Meeting the Recovered Xichang Street —— Joint Graduation Project of Four Schools

Instructor: Shen Yao

Student: Guo Xiaokang

● 设计说明

如果西长街是一个生命体，那么它正饱受着各种病痛的折磨，如发热、眼疾、耳疾、呼吸病、肌无力、身体不协调等。而持续数月的新冠肺炎疫情更是给其雪上加霜，后疫情时代的西长街何去何从？

健：为西长街治病，带动街区机体的康复和痊愈，减少未来可能的健康隐患是第一层目标。

趣：为西长街疗心，提高具体部位的活力和弹性，促进活动触媒的发生演进是第二层目标。

Design notes

If Xichang street is a living body, it is suffering from all kinds of diseases. Such as fever, eye diseases, ear diseases , respiratory diseases , myasthenia physical disharmony, etc. One disaster after another is COVID-19. Decide on what path to follow in the post epidemic era? Where is it going?

Health: the first goal is to treat diseases in Xichang street, drive the recovery of the block body, and reduce possible health hazards in the future.

Interest: the second goal is to heal the heart for Xichang street, improve the vitality and elasticity of specific parts, and promote the occurrence and evolution of activity catalysts.

● 健康专题研究 Special research on health

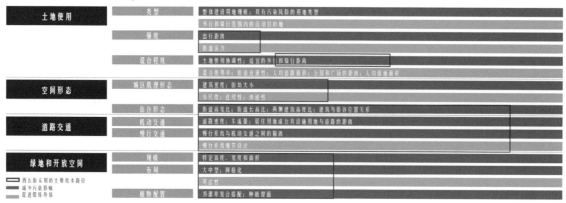

● 设计导则 Design guidelines

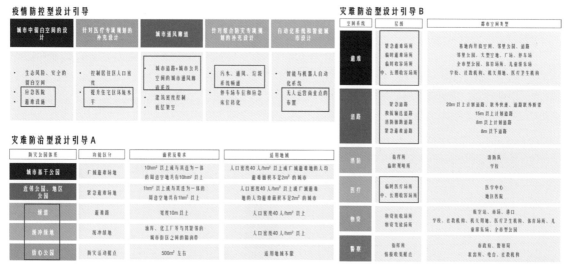

● 区位概述 Location overview

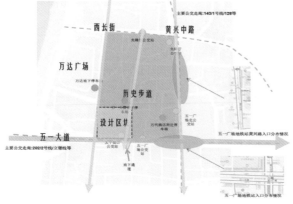

外部交通发达

场地整体感受欠佳，环境品质有待提升

● 环境评价 Environmental assessment

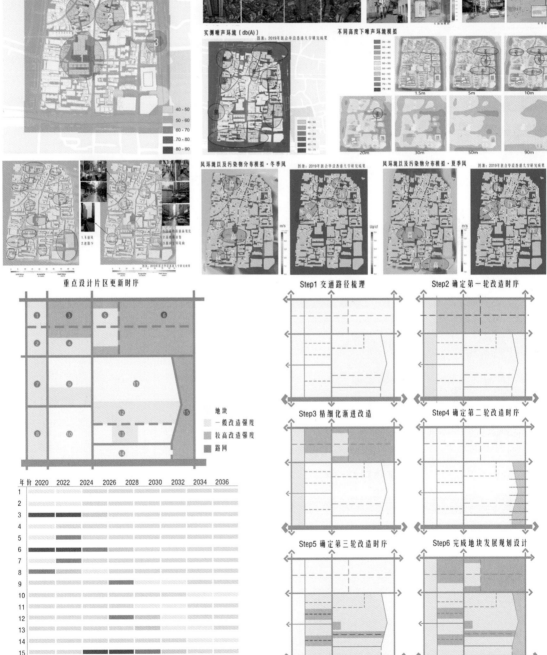

图源：2019年联合华设香港大学研究成果

实测噪声环境（db(A)）

不同高度下噪声环境模拟

风环境及及污染物分布模拟·冬季风　　风环境及及污染物分布模拟·夏季风

重点设计片区更新时序

各个地块更新时序示意图

Step1 交通路径梳理　　Step2 确定第一轮改造时序

Step3 精细化渐进改造　　Step4 确定第二轮改造时序

Step5 确定第三轮改造时序　　Step6 完成地块发展规划设计

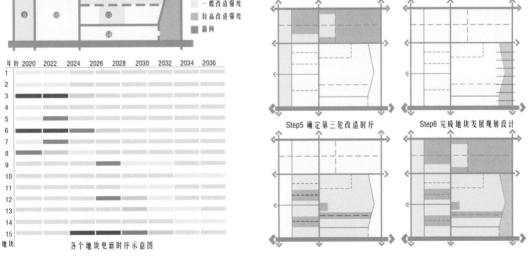

● 片区构想　Regional concept

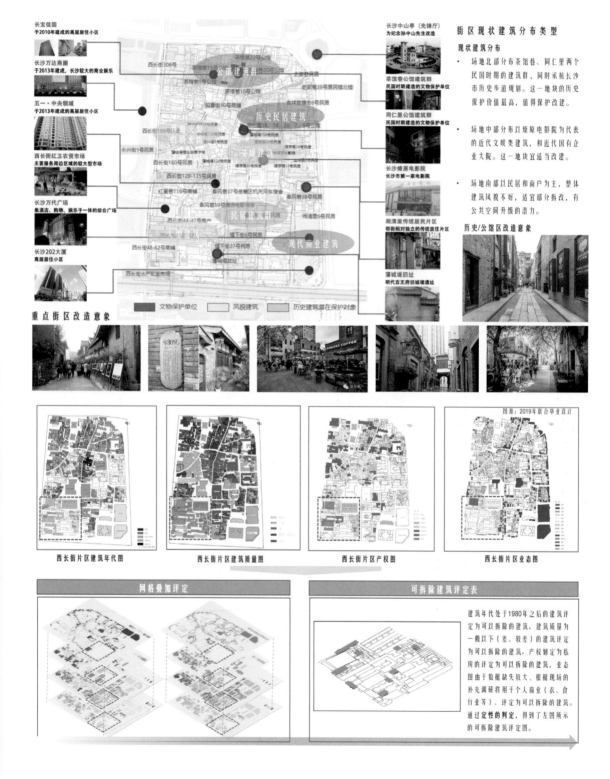

规划设计　Planning and design

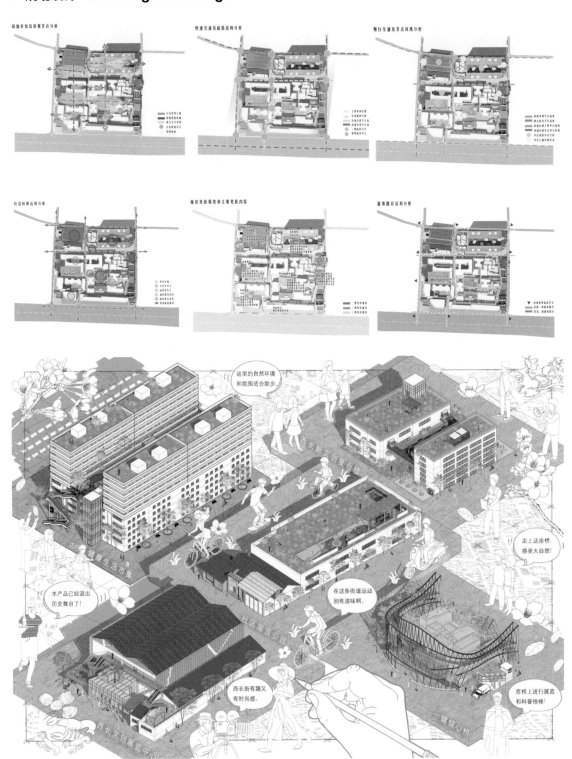

文 创 展 览 中 心 界 面

五 一 大 道 临 街 界 面

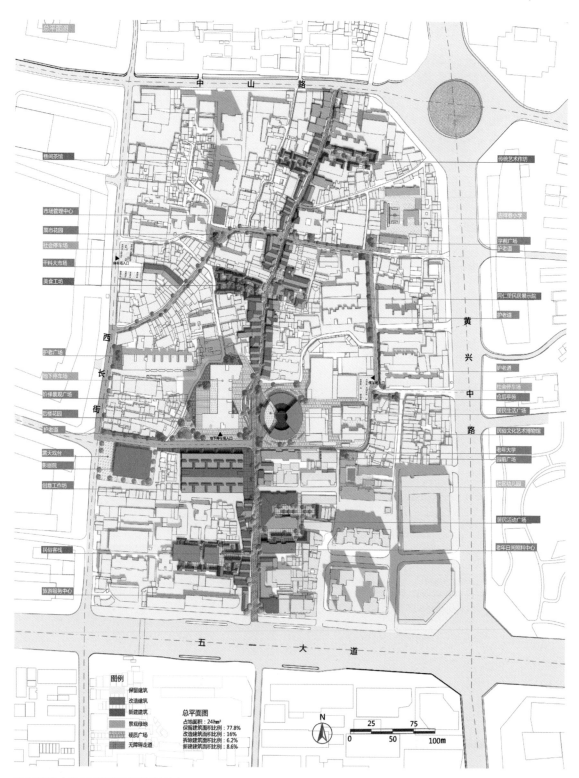

图例
- 保留建筑
- 改造建筑
- 新建建筑
- 景观绿地
- 硬质广场
- 无障碍走道

中 山 路

巷间茶馆

市场管理中心
菜市花园
社会停车场
干料大市场
美食工坊

护老广场
地下停车场
阶梯景观广场
后楼花园
护老道

露天戏台
影剧院
创意工作坊

民俗客栈

旅游服务中心

西 长 街

传统艺术作坊

吉洋巷小学

学前广场
护老道

同仁里民居展示院
护老道

社会停车场
仓后亭苑
居民生活广场

民俗文化艺术博物馆

老年大学
园前广场

社区幼儿园

居民活动广场

老年日间照料中心

黄 兴 中 路

五 一 大 道

总平面图
占地面积：24 hm²
保留建筑面积比例：77.8%
改造建筑面积比例：16%
拆除建筑面积比例：6.2%
新建建筑面积比例：8.6%

N

0 25 50 75 100m

总平面图（老年友好旧城更新设计）

176

设计题目: 西长街历史步道片区老年友好旧城更新设计
指导老师: 沈瑶
学　　生: 姜开源

Design topic: The Renewal Design of the Elderly-Friendly Old City in the Historical Trail Area of XiChang Street

Instructor: Shen Yao

Student: Jiang Kaiyuan

● 设计说明

本次毕业设计主题为长沙市历史步道两厢历史城区西长街片区的更新详细规划,规划研究分析范围为长沙历史步道城区范围。西长街片区现状建筑质量差,范围内有数量很多的具有历史价值的建筑,居住环境差,居住其中的大多为流动人口即租户,老年人比例很大。设计重点是对片区内"老人友好"的空间氛围进行打造。

Design notes

The theme of this graduation design is the detailed planning of the Xichang Street area of the historical district of the historical trail in Changsha City. The research and analysis area covered the urban area of the historical trail of Changsha. The current construction quality in the Xichang Street area is poor. There are many buildings with historical value in the area. The living environment is poor. Most of the residents are migrant population or tenants, and the proportion of elderly people is large. The focus of the design is to create an "elderly friendly" space atmosphere in the area.

● 轴线分析　Axis analysis

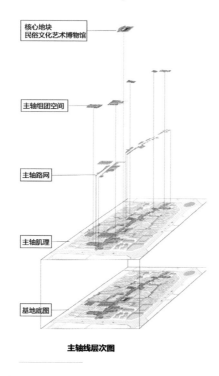

核心地块
民俗文化艺术博物馆

主轴组团空间

主轴路网

主轴肌理

基地底图

主轴线层次图

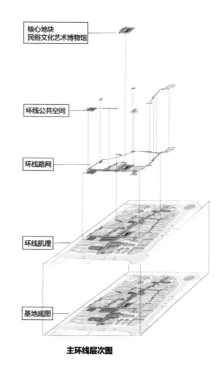

核心地块
民俗文化艺术博物馆

环线公共空间

环线路网

环线肌理

基地底图

主环线层次图

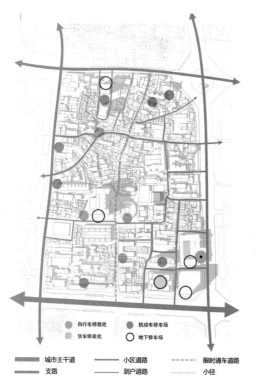

自行车停放处　　机动车停车场
货车停放处　　地下停车场

━━ 城市主干道　　━ 小区道路　　┈┈ 限时通车道路
━ 支路　　━ 到户道路　　┈┈┈ 小径

┈ 社区10分钟步行圈
◀┈┈▶ 各步行圈步行联系路径

178

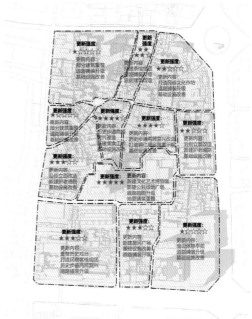

★ 数量越多，更新强度越大，更新越早

景观轴　　公共绿地（口袋公园）　　社区绿地

茶馆

居民生活广场

传统民宿区

仓后亭苑

179

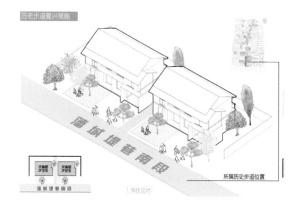

作为城市规划者、城市设计师，要在让老人拥有更好的生活品质和生活环境的同时，也要让他们感受到自己的社会认同感，让他们有机会、有条件发挥他们的价值，这也是本设计的主旨，即"老有所为"。

As urban planners and urban designers, we should not only let the elderly have a better quality of life and living environment, but also let them feel their social identity, so that they have the opportunity and conditions to give full play to their value. This is also the main purpose of this design, that is, to "provide the elderly with a sense of worthiness".

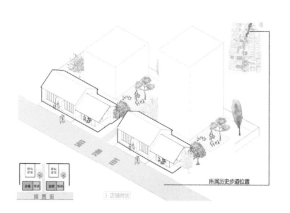

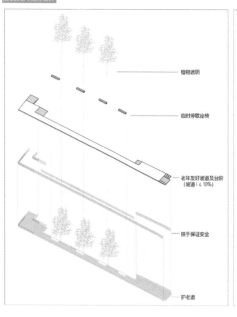

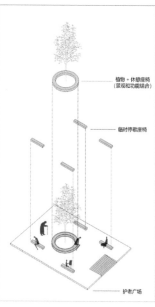

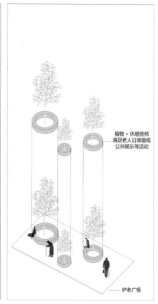

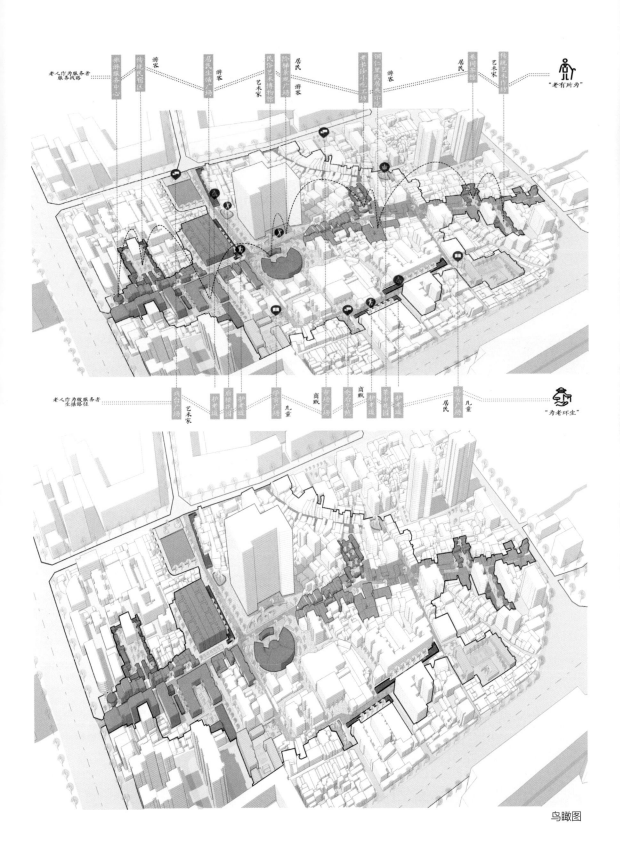

老人作为被服务者
服务线路

老人作为被服务者
生活路径

鸟瞰图

181

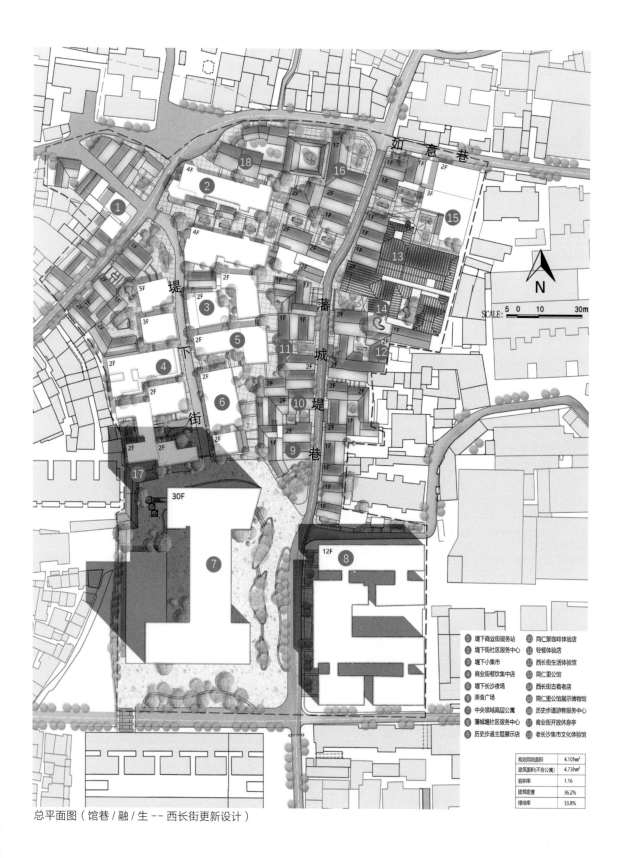

如 意 巷

堤 下 街

潘 城 堤 巷

N

SCALE: 5 0   10   30m

规划用地面积	4.10 hm²
建筑面积(不含公寓)	4.73 hm²
容积率	1.16
建筑密度	36.2%
绿地率	33.8%

❶ 堤下商业街服务站　　❿ 同仁聚咖啡体验店
❷ 堤下街社区服务中心　⓫ 轻餐体验店
❸ 堤下小集市　　　　　⓬ 西长街生活体验馆
❹ 商业街餐饮集中店　　⓭ 同仁里公馆
❺ 堤下长沙夜场　　　　⓮ 西长街古看老店
❻ 美食广场　　　　　　⓯ 同仁里公馆展示博物馆
❼ 中央领域高层公寓　　⓰ 历史步道游客服务中心
❽ 潘城堤社区服务中心　⓱ 商业街开放休息亭
❾ 历史步道主题展示店　⓲ 老长沙集市文化体验馆

总平面图（馆巷／融／生－－西长街更新设计）

设计题目： 馆巷 / 融 / 生——西长街更新设计
指导老师： 沈瑶
学　　生： 陈彦伊

Design topic: Pavilion Lane/Integration/Rebirth——Renewal Design of Xichang Street

Instructor: Shen Yao

Student: Chen Yanyi

● **设计说明**

对长沙市开福区西长街历史片区从空间人群社会活动等方面进行调查研究，挖掘长沙旧城城市发展逻辑以及空间肌理变化，结合开福区历史步道上位规划政策对西长街片区进行空间解读，根据调研得出场地的综合发展特质，以指导旧城更新设计。

Design notes

Investigate and research the historical area of Xichang Street in Kaifu District, Changsha from the social activities of people, and explore the urban development logic and spatial texture changes of the old city of Changsha. Combine the upper planning policy of the historical trail in Kaifu District to interpret the space of the Xichang Street area. The comprehensive development characteristics of the site are obtained based on the research to guide the renewal design of the old city.

● 旧城更新规划主题阐述　Description of the theme of the old city renewal planning

旧城现状发展总结

一方面，老建筑的拆除和改造，伴随着无序生长的新建筑，让老城区传统的风貌逐渐改变。传统建筑语汇、空间尺度、历史肌理和社区特征也被逐渐弱化。另一方面，随着建筑和邻里社区的老化以及缺乏及时修缮，老城区整体的环境已不能满足现代生活品质的要求，导致老城区本地人口不断减少，年轻人流失严重，城市的活力也在逐渐消逝。

场地突出感受——元素"混合"

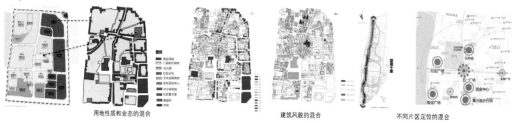

用地性质和业态的混合　　　　建筑风貌的混合　　　　不同片区定位的混合

规划主题概念

人群共享与
功能混合的
"馆巷新生"

● 片区概念规划　Conceptual planning of the district

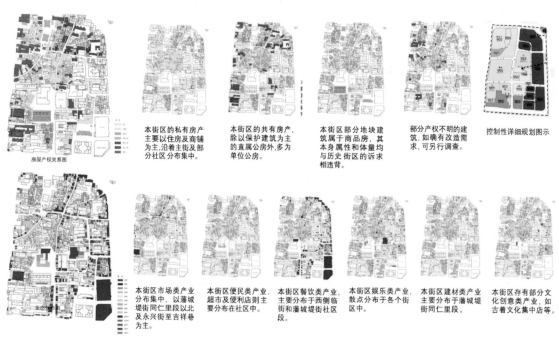

房屋产权关系图

本街区的私有房产，主要以住房及商铺为主，沿着主街及部分社区分布集中。

本街区的共有房产，除以保护建筑为主的直属公房外，多为单位公房。

本街区部分地块建筑属于商品房，其本身属性和体量均与历史街区的诉求相违背。

部分产权不明的建筑，如确有改造需求，可另行调查。

控制性详细规划图示

业态图

本街区市场类产业分布集中，以藩城堤街同仁里段以北及永兴街至吉祥巷为主。

本街区便民类产业，超市及便利店则主要分布在社区中。

本街区餐饮类产业，主要分布于西侧临街和藩城堤街社区段。

本街区娱乐类产业，散点分布于各个街区中。

本街区建材类产业，主要分布于藩城堤街同仁里段。

本街区存有部分文化创意类产业，如古着文化集中店等。

● "拆改留" 示意图 Schematic diagram of demolition,replacement and retention

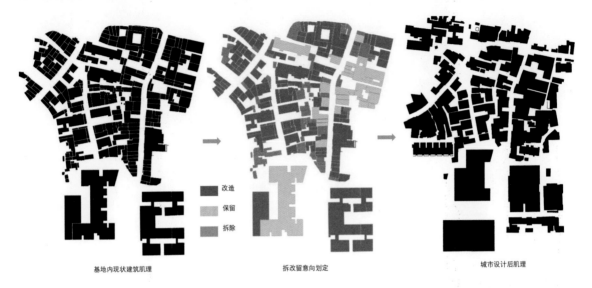

改造
保留
拆除

基地内现状建筑肌理   拆改留意向划定   城市设计后肌理

● 设计分析 Diagram

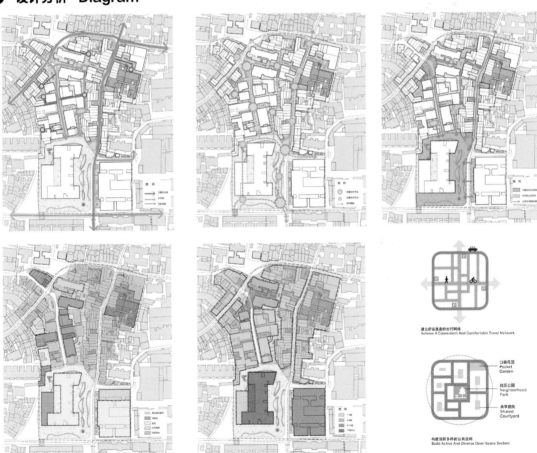

建立舒适通畅的出行网络
Achieve A Convenient And Comfortable Travel Network

口袋花园
Pocket Garden
社区公园
Neighborhood Park
共享庭院
Shared Courtyard

构建活跃多样的公共空间
Build Active And Diverse Open Space System

● 沿街立面　Facade along the street

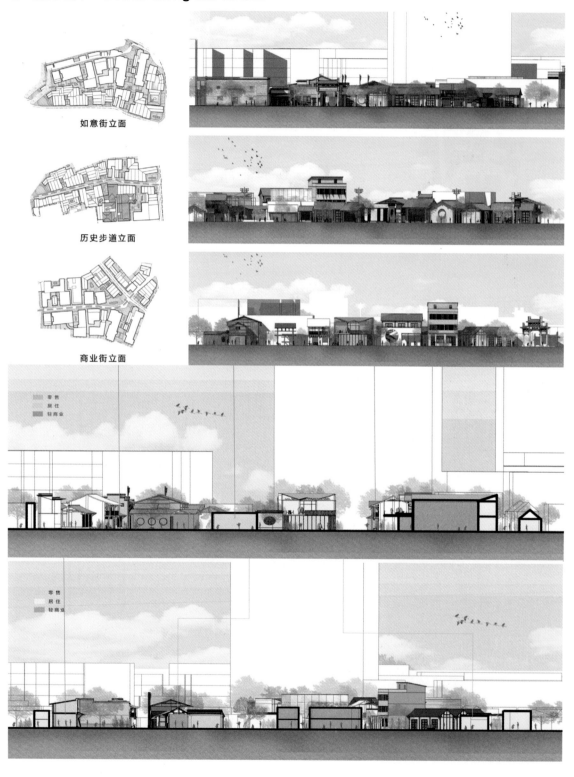

如意街立面

历史步道立面

商业街立面

vintage风格哦,店里还挺热闹,以后有空可以常来啦!

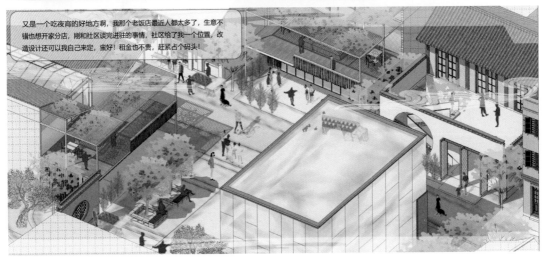

又是一个吃夜宵的好地方啊,我那个老饭店最近人都太多了,生意不错也想开家分店,刚和社区谈完进驻的事情,社区给了我一个位置,改造设计还可以我自己来定,蜜好!租金也不贵,赶紧占个码头!

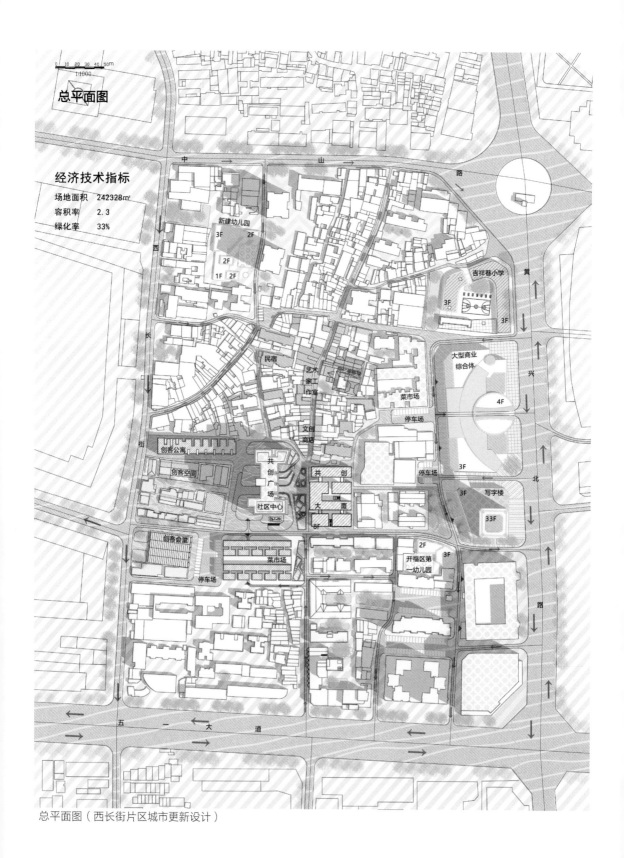

总平面图

经济技术指标

场地面积	242328m²
容积率	2.3
绿化率	33%

新建幼儿园
3F 2F
2F
1F 2F

吉祥巷小学
3F
3F

民宿
艺术家工作室
菜市场
停车场

大型商业综合体
4F

文创商店
3F

创客公寓
创客空间
共创广场
社区中心
共创大厦
8F

共创广场
停车场
3F

写字楼
3F
33F

创客会堂
停车场
菜市场

2F
开福区第一幼儿园
3F

中 山 路
西 长 街
黄 兴 北 路
五 一 大 道

总平面图（西长街片区城市更新设计）

188

设计题目： 长沙市西长街片区城市更新规划设计
指导老师： 沈瑶
学　　生： 杨乐川

15

Design topic: Urban Renewal Planning and Design of Xichang Street Area in Changsha City

Instructor: Shen Yao

Student: Yang Lechuan

● 设计说明

本设计的重点是匹配空间与人群的适应性，旨在实现西长街街区的宜居、繁荣和可持续的发展目标。首先通过设计便捷的交通、配备足够的公共服务设施和基础设施来打造宜居的生活环境。其次通过引入丰富的业态，满足原住民、租户、艺术家、游客以及市民等不同人群的活动需求。

Design notes

The focus of this design is to match the adaptability of Space and People, in order to achieve the livable, prosperous and sustainable development goal of Xichang Street Area. First, A livable Living Environment is created by designing Convenient Transportation and equipped with Sufficient Public Service Facilities and Infrastructure. Secondly, through the introduction of various Business Forms, we can meet the activity needs of different groups such as Aborigines, Tenants, Artists, Tourists and Citizens.

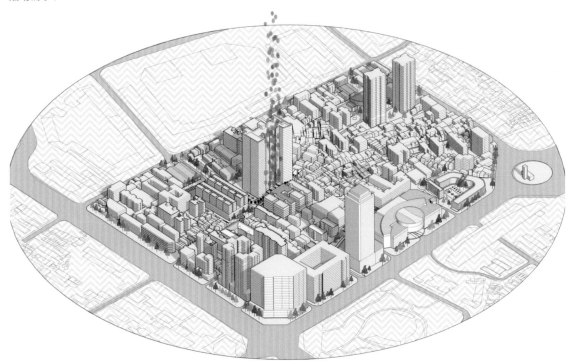

● 场地区位　Field location

● 现状分析　Current state analysis

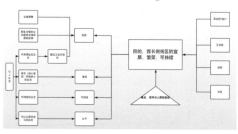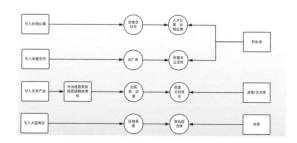

● 方案逻辑　Scheme logic

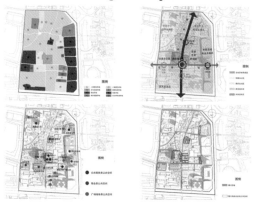

● 设计分析　Design analysis

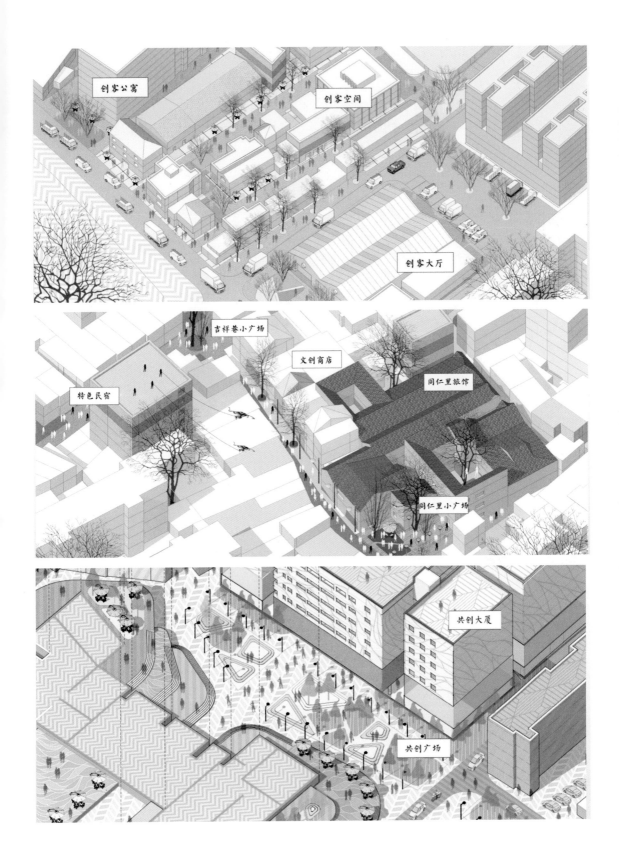

创客公寓　创客空间　创客大厅

特色民宿　吉祥巷小广场　文创商店　同仁里旅馆　同仁里小广场

共创大厦　共创广场

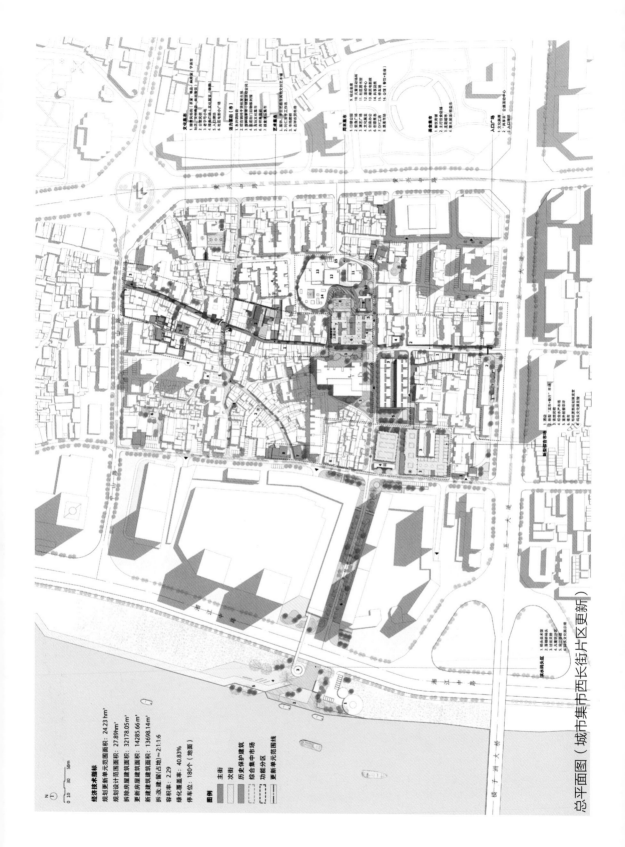

总平面图（城市集市西长街片区更新）

经济技术指标

规划更新单元范围图面积：24.23 hm²

规划设计范围图面积：27.89hm²

拆除房屋建筑面积：32178.05m²

更新房屋建筑面积：14285.66 m²

新建建筑建筑面积：13698.14m²

拆改建量（占地）≈2:1:6

容积率：2.29

绿化覆盖率：40.83%

停车位：180个（地面）

图例

主街

次街

历史保护建筑

综合集中市场

功能分区

更新单元范围线

16

设计题目： 西长街片区更新设计——城市集市

指导老师： 沈瑶

学　　生： 廖鑫

Design topic: Renewal Design of Xichang Street Area —— Urban Market

Instructor: Shen Yao

Student: Liao Xin

● 设计说明

在本地人的印象里，西长街就是有名的水产海鲜街，人们对菜市场的认知比对这整个场地的认知度高，访谈中大家多多少少都会提及对于菜市的看法，虽然它给予了很大的方便，但是卫生环境还是它最大的问题。

人们对于这块场地产生了感情，尤其是在这里生活了十几年的老住户，又爱又恨，但爱大于恨。很多老年人甚至有了依赖感，即便搬离了西长街，还是想回到这个最熟悉的地方买菜。

"菜市场"成了人们日常生活中的必经之地，家长下班后接了小孩去买菜，老人平时到街上聊聊天，于是，菜街便成了西长街片区人们交往生活的必经之路。

Design notes

In the impression of local people, Xichang Street is a famous aquatic and seafood street. The cognition of the vegetable market is higher than that of the whole site. In the interview, everyone will mention their views on the vegetable market. Although it has given great convenience, the sanitary environment is still its biggest problem.

People have feelings for this site, especially the old residents who have lived here for more than ten years. They love and hate it but love is greater than hate. Many old people even have a sense of dependence. Even if they move away from Xichang Street, they still want to go back to this most familiar place to buy vegetables.

"Vegetable market" has become a necessary place in people's daily life. Parents pick up children to buy vegetables after work. When the old man is bored, he goes to the street to chat... Vegetable street has become the only way for people to communicate and live in Xichang street.

● 专题调研 Special research

- 01:00 商贩不紧不慢地在撕箱子、拿秤、接电线，一直忙活了半个多小时，才把他们一晚上要兜售的几大箱小龙虾全搬到了马路边。
- 02:00 很多自制的小灯泡纷纷在街边亮起，室成了星星点点的街灯。很多顾客已经骑着摩托、踏着三轮、开着小皮卡来这里采购了。
- 04:00 很多商家都选择当街宰杀，空气中弥漫着一股浓重的血腥味、鱼腥味。西长街上多了很多摆着泡沫箱子在街边卖海鲜水产的商贩。五一大道上停着很多面包车，吉祥巷菜市的很多商贩也开始推着三轮车，骑着电动车往返于大街上进菜。
- 05:00 菜市场在自家店铺门口摆菜，很多餐馆老板开始来菜市采购上午食材。
- 06:00 早餐店陆陆续续开门了，大爷大妈锻炼完来市场买菜。
- 07:00 菜贩给蔬菜洒水，肉铺开始剁肉——迎接买菜早高峰。
- 09:00 买菜的人群逐渐变少，其他店铺（裁缝店、外卖店、冷作店）开始营业。
- 11:00 菜市场又一次经历一波买菜的高峰。很多外卖小哥骑着电动车快速往返于市场中间，餐饮店生意格外火爆。
- 12:00 菜市场逐渐进入午休，菜贩们坐在椅子上打瞌睡。
- 15:00 开始给蔬菜补洒水，新一轮摘菜，为饭店晚饭时间段做准备。
- 17:00 菜贩将食材运达指定饭店。
- 18:00 下午买菜的高峰，路段十分拥堵。街道上聚集了下班回家、接小孩、遛狗买菜等各种人群，菜街生意特别好，买菜的主要以中青年为主，有很多推三轮车的流动铺贩，肉禽的腥臭味特别明显，地面非常脏，堆积了一整天的垃圾和污水。
- 19:00 买菜的人群逐渐变少，菜贩抓紧时间卖菜边吃饭。街巷里你望着两旁居民楼里散发出来的饭菜香，商店陆陆续续关门，菜贩收摊。
- 20:00 天已经全黑，橙黄色的灯光亮起来 巷子里零零星星有出来散步的居民，沿主街的餐饮店生意十分红火。
- 24:00 沿街夜宵摊比较热闹，西长街内格外安静，正等待着崭新的一天来临。

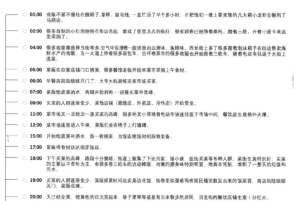

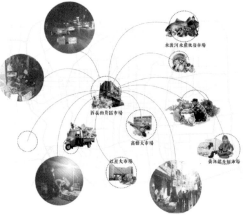

| 1000 | 250 | 2500 | 250 | 1000 |

吉祥巷

| 1000 | 250 | 3000 | 250 | 800 |

潘城路巷

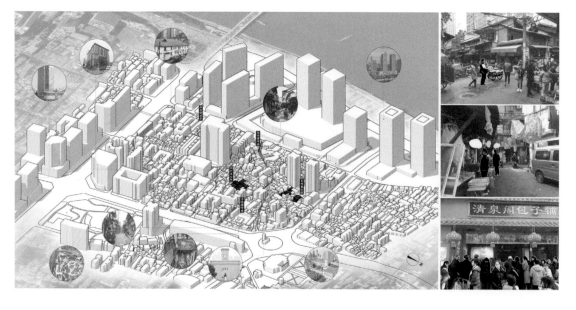

● 概念分析　Conceptual analysis

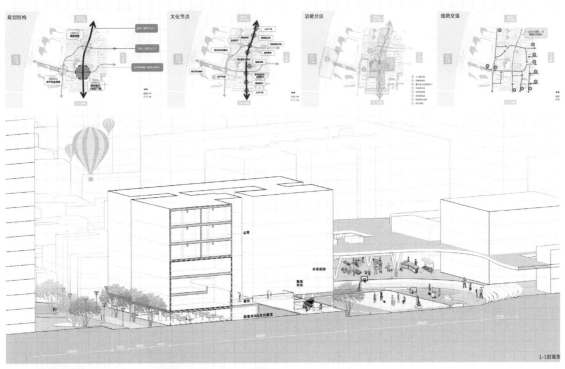

1-1剖面图

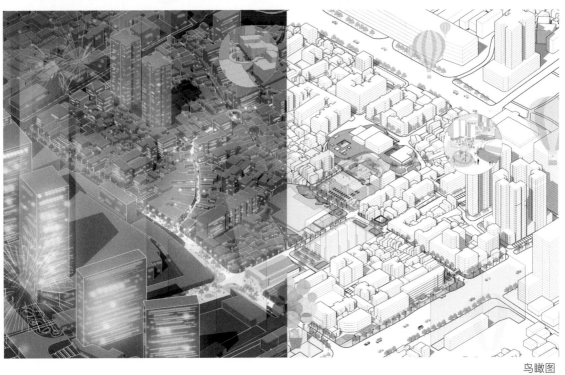

鸟瞰图

新型现代市场
1.合理的动线布局与功能分区
2.大型集中，干净整洁，管理有序
3."菜场+餐厅"
4.人群年轻化

集市
1.主题策划，集市定位
2.时间更长，覆盖整块场地，白天夜晚的转换
3.现代年轻，时尚艺术

商业街
1.业态提质，更符合城市中心地段
2.步行友好，业态丰富，吸引更多人群

历史步道
1.城市文化名片
2.结合"市"形成独有特色，打造最有市井味道的街区

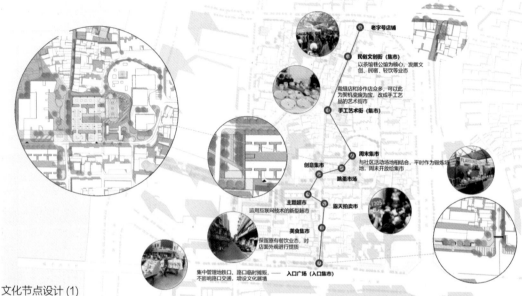

文化节点设计 (1)

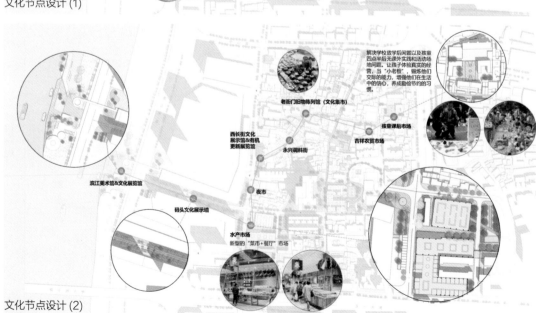

文化节点设计 (2)

● 建筑更新改造与节点　Building renovation and node

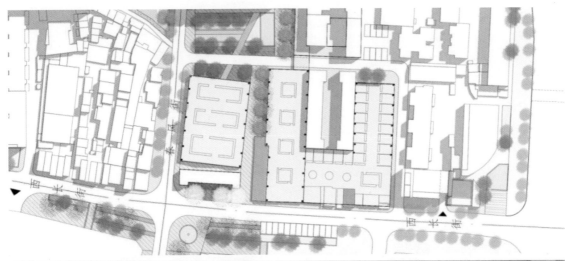

新型综合水产市场

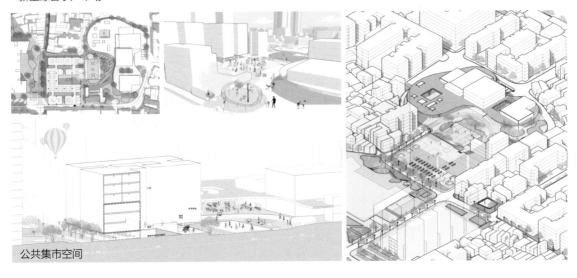

公共集市空间

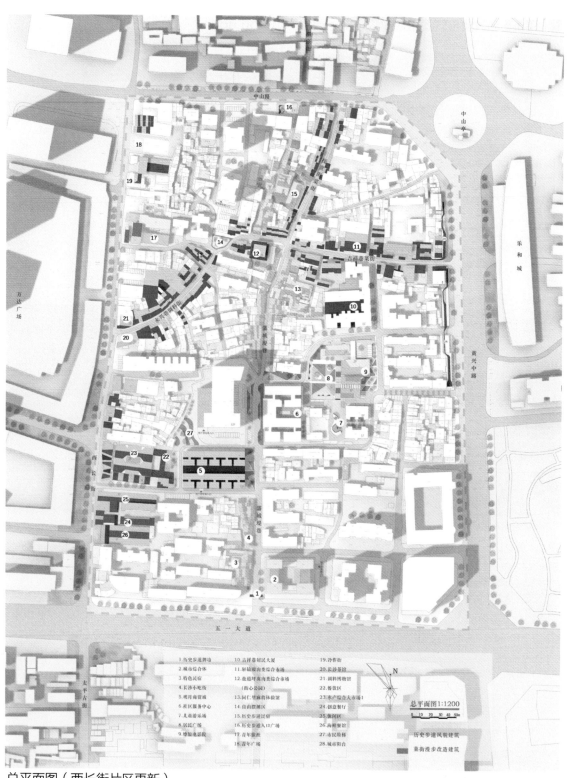

总平面图（西长街片区更新）

设计题目：历史城区西长街片区的更新设计
指导老师：沈瑶
学　　生：浦钰婷

17

Design topic: Renewal Design of the Historical Area—Xichang Street

Instructor: Shen Yao

Student: Pu Yuting

● 设计说明

本次毕业设计主题为长沙市历史步道两厢历史城区西长街片区的更新详细规划。规划研究分析范围为长沙历史步道城区范围（西至湘江中路、东至芙蓉路、北至三一大道、南至白沙路，面积约 8.1 km²），包括历史城区（北至湘春路、东南到芙蓉路、建湘路、白沙路、西至湘江，总用地面积为 5.6km²，为明清长沙老城范围）。本次设计试图在历史步道街区更新下以菜市场的改造与升级为契机，为市民构建城市宜居生活圈，打造长沙市井生活的旅游名片 ---- 城市菜街漫步。

Design notes

The theme of this graduation project is the updated detailed planning of Xichang Street area of Liangxiang historical urban area of Changsha historical footpath. The scope of planning research and analysis is the urban area of Changsha historical footpath (middle Xiangjiang Road in the west, Furong Road in the East, Sany Avenue in the north and Baisha road in the south, covering an area of about 8.1 square kilometers), including the historical urban area (Xiangchun road in the north, Furong Road in the southeast, Jianxiang Road, Baisha road in the southeast and Xiangjiang River in the west, with a total land area of 5.6 square kilometers, which is the old city of Changsha in the Ming and Qing Dynasties). This paper attempts to build an urban livable life circle for citizens and create a tourism card of Changsha City Life – urban vegetable street walk under the renewal of historical footpath blocks and the opportunity of the transformation and upgrading of vegetable market.

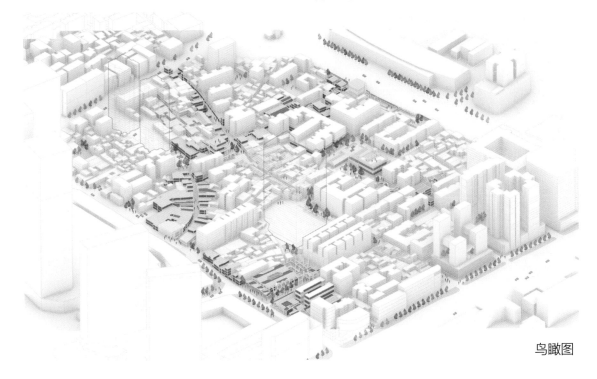

鸟瞰图

● 概念生成　Concept generation

从菜市场中选择改造节点，以节点带动菜街与社区活力，以菜街为漫步轴线，打造特色片区；结合历史步道，联动周边片区，为不同人群提供公共活动的空间，打造集日常与旅游于一体的菜街漫步体系。

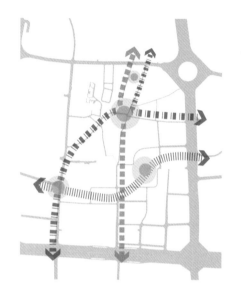

设计分析

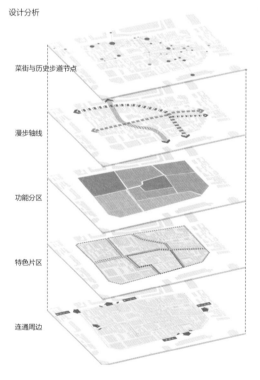

菜街与历史步道节点

漫步轴线

功能分区

特色片区

连通周边

● 业态更新　Business renewal

（1）相互辅助促进：根据不同的业态特殊的需求，以及市民方便就近的购菜路线，布置合理的菜市场业态；
（2）街道整洁；
（3）结合历史步道优化旅游路线；
（4）增加商户自治管理委员会。

● 更新机制　Update mechanism

● 菜街更新　Vegetable street renewal

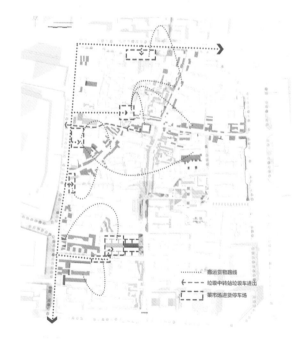

● 风貌改造　Appearance transformation

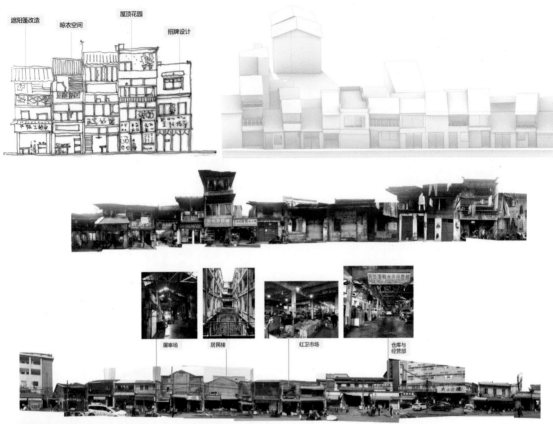

遮阳篷改造　晾衣空间　屋顶花园　招牌设计

屠宰场　居民楼　红卫市场　仓库与经营部

● 西长街水产综合片区　Xichang street aquatic products comprehensive area

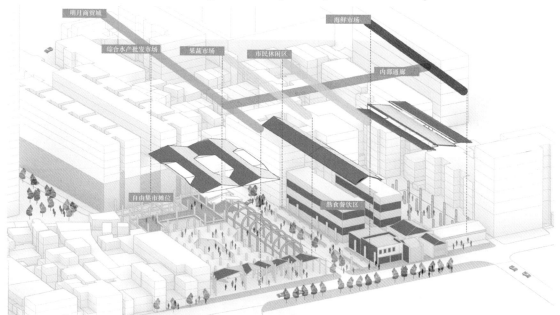

● 街心公园设计　Street park design

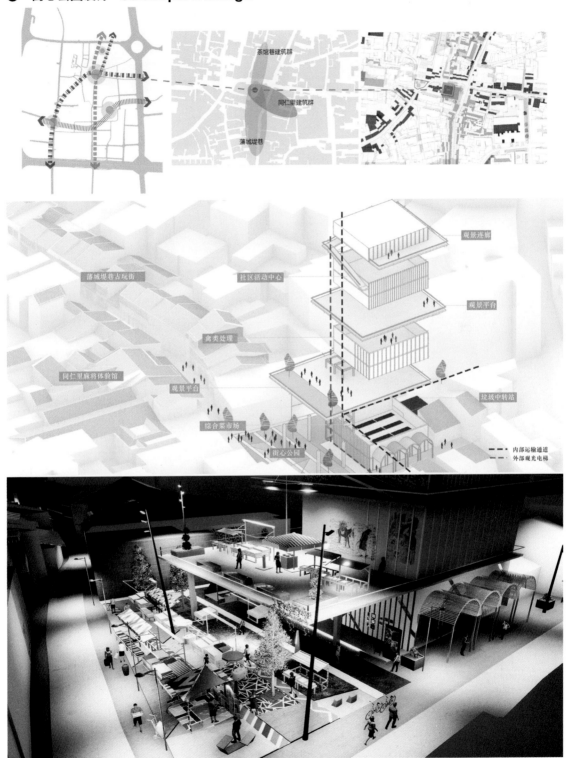

观景连廊

观景平台

垃圾中转站

社区活动中心

藩城堤巷古玩街

禽类处理

同仁里麻将体验馆

观景平台

综合菜市场

街心公园

内部运输通道
外部观光电梯

茶馆巷建筑群

同仁里建筑群

藩城堤巷

总平面图

18

设计题目：立异创新 相融共生——潮宗街历史区城市更新设计
指导老师：向辉　孙亮
学　　生：仁安之

Design topic: Innovation, Integration and Symbiosis —— Urban Renewal Design of Chaozong Street Historical Area

Instructors: Xiang Hui,Sun Liang

Student: Ren Anzhi

● 设计说明

基地位于长沙市 6 个分区之一的芙蓉区，处于长沙市主城东部，交通便利。芙蓉区是浓缩长沙城市发展历程的代表性城区，聚集了彰显长沙历史文化底蕴的相当大一部分精华。同时基地位处长沙市中心五一广场商圈的西北角，与市中心直线距离仅为 1km 左右。基地周边有三条主干道以及一条城市次干路，地铁一号线在场地东侧穿过，北边为培元桥地铁站。场地区位优势明显。

Design notes

The base is located in Furong district, one of the six districts of Changsha, and is located in the east of the main city of Changsha with convenient transportation. The Furong district is a typical urban area that concentrates the development of Changsha's city. It has gathered a large part of the essence of Changsha's historical and cultural heritage. At the same time, the base is located in the northwest corner of Wuyi Square business circle in the center of Changsha, which is only about 1km away from the center of the city. There are three main roads and one urban secondary road around the base. Metro Line 1 passes through the east of the site, and peiyuanqiao metro station is in the north corner. It has obvious location advantages.

立异创新 相融共生 和而不同 美美与共

● 方案生成过程　Generation process

1.板块划分，功能定位　　2.结合上位规划，明确保整拆　　3.确定关键节点，释放空间　　4.设计人车流线，串联节点　　5.节点空间深化

● 功能植入　Functional implantation

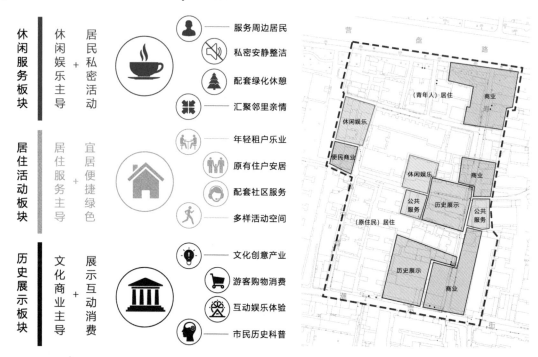

休闲服务板块	休闲娱乐主导 +	居民私密活动	服务周边居民 / 私密安静整洁 / 配套绿化休憩 / 汇聚邻里亲情
居住活动板块	居住服务主导 +	宜居便捷绿色	年轻租户乐业 / 原有住户安居 / 配套社区服务 / 多样活动空间
历史展示板块	文化商业主导 +	展示互动消费	文化创意产业 / 游客购物消费 / 互动娱乐体验 / 市民历史科普

● 流线重组　Streamline reorganization

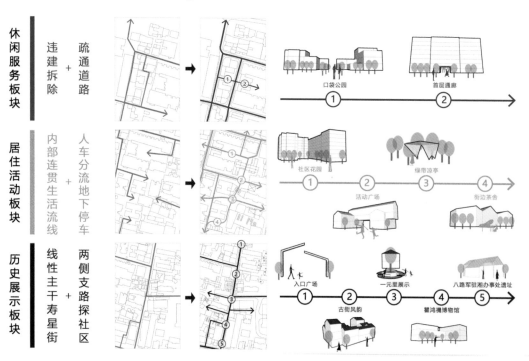

休闲服务板块	违建拆除 + 疏通道路		口袋公园 ① / 首层通廊 ②
居住活动板块	内部连贯生活流线 + 人车分流地下停车		社区花园 ① / 活动广场 ② / 绿带凉亭 ③ / 街边茶舍 ④
历史展示板块	线性主干寿星街 + 两侧支路探社区		入口广场 ① / 古街风韵 ② / 一元里展示 ③ / 蟫鸿榭博物馆 ④ / 八路军驻湘办事处遗址 ⑤

● 板块路径连接　Path connection

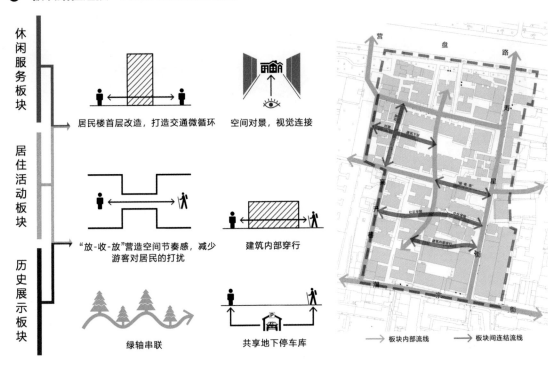

休闲服务板块

居住活动板块

历史展示板块

居民楼首层改造，打造交通微循环

空间对景，视觉连接

"放-收-放"营造空间节奏感，减少游客对居民的打扰

建筑内部穿行

绿轴串联

共享地下停车库

→ 板块内部流线　　→ 板块间连结流线

● 风貌提质　Style improvement

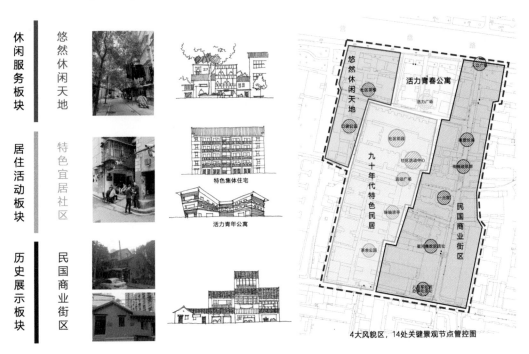

休闲服务板块　悠然休闲天地

居住活动板块　特色宜居社区

历史展示板块　民国商业街区

特色集体住宅

活力青年公寓

悠然休闲天地

活力青春公寓

九十年代特色民居

民国商业街区

4大风貌区，14处关键景观节点管控图

● 空间特色分析　Spatial feature analysis

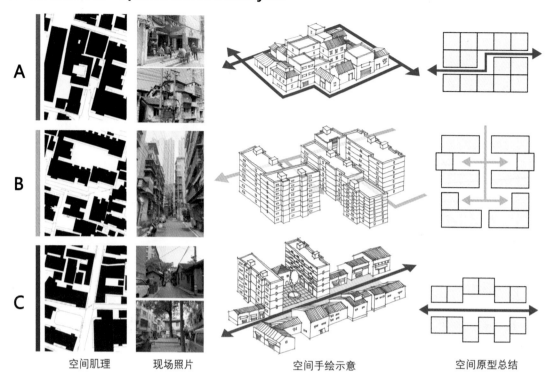

A

B

C

空间肌理　　　现场照片　　　　　空间手绘示意　　　　空间原型总结

● 空间活化策略　Spatial activation strategy

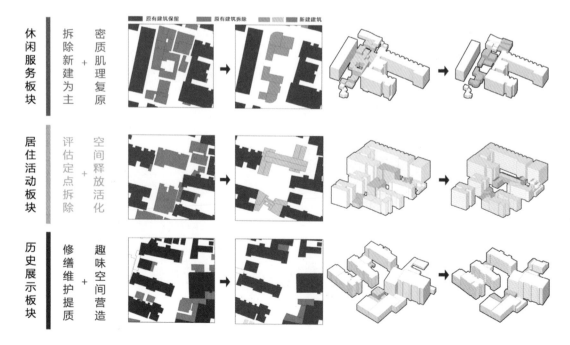

休闲服务板块　拆除新建为主＋密质肌理复原

居住活动板块　评估定点拆除＋空间释放活化

历史展示板块　修缮维护提质＋趣味空间营造

原有建筑保留　　原有建筑拆除　　　新建建筑

● 空间设计　Space design

改造前——居住质量欠佳

改造后——改善青年租户条件

改造前——违建堵塞道路

改造后——密路网，尺度宜人

改造前——阻断，低效利用空间

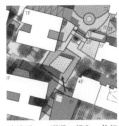

改造后——联通，绿色，休闲

改造前——隔离，封闭

改造后——联动，激活

改造前——缺少纪念意义

改造后——历史教育，文化展示

改造后（穿过建筑首层到达悦动广场）

改造后（穿过社区活动中心到达活动广场）

改造后（从街边望向瞿鸿禨博物馆）

改造后（从高升巷由南向北望）

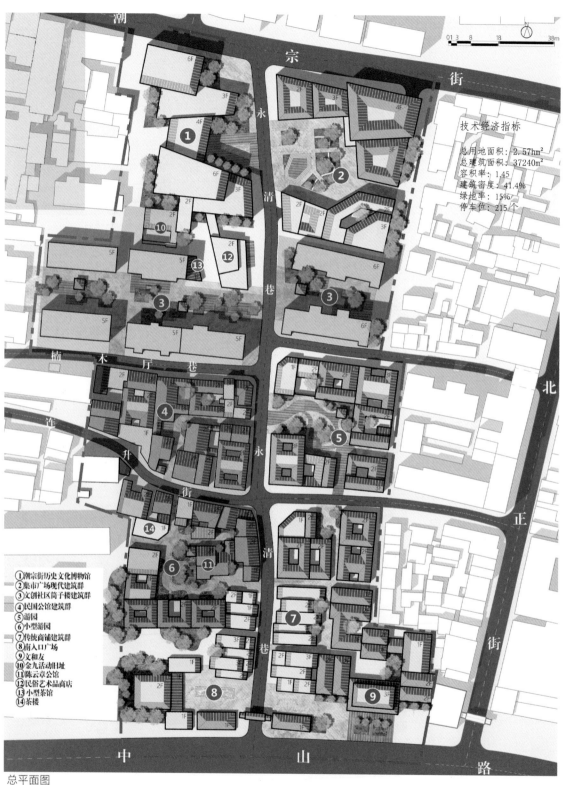

潮　宗　　　街

永
清
巷

北

楠　木　厅　巷

连
升
街

永
清
巷

正

街

中　　　　山　　　　路

01 3 8   18   38m

技术经济指标

总用地面积：2.57hm²
总建筑面积：37240m²
容积率：1.45
建筑密度：41.4%
绿地率：15%
停车位：215个

①潮宗街历史文化博物馆
②集市广场现代建筑群
③文创社区筒子楼建筑群
④民国公馆建筑群
⑤游园
⑥小型游园
⑦传统商铺建筑群
⑧南入口广场
⑨文和友
⑩金九活动旧址
⑪陈云章公馆
⑫民俗艺术品商店
⑬小型茶馆
⑭茶楼

总平面图

210

设计题目： 阅读历史 触摸时光——永清巷片区历史保护规划设计

指导老师： 向辉 孙亮

学 生： 张达

19

Design topic: Reading History and Feeling Time —— Planning and Design of Historical Protection of Yongqing Lane

Instructors: Xiang Hui，Sun Liang

Student: Zhang Da

● 设计说明

潮宗街历史文化街区位于长沙市湘江以东地区，地处长沙市中心区域北部，与南部的西长街历史文化街区、太平街历史文化街区共同构成了长沙市历史步道的核心部分。潮宗街历史文化街区西部为湘江中路，北部为营盘路，东部为黄兴中路，南部为中山路，总面积约为20.86hm²。

潮宗街历史文化街区内历史街道众多，除最主要的潮宗街外，还包含北正街、连升街、福庆街、永清巷、楠木厅巷等历史文化氛围较为浓厚的街道；其中潮宗街与北正街同太平老街一样，为长沙市重点历史文化保护项目。潮宗街东起湘江路，西止黄兴北路，是长沙城仅存的4条古麻石街之一，今存长约400m，宽9m，为旧时最宽的街道。潮宗街历史文化街区地理位置较为优越，周边有较多长沙市著名景点或历史建筑，如万达广场、长沙少年宫、五一广场以及中山亭、北正街基督堂、西长街民居建筑群等历史建筑。

Design notes

Chaozong Street Historical and Cultural Block is located in the east of Xiangjiang River in Changsha, in the north of Changsha's downtown area. It forms the core part of Changsha's historical trails together with the Xichang Street Historical and Cultural Block and Taiping Street Historical and Cultural Block in the south.The Chaozong Street Historical and Cultural Block has Xiangjiang Middle Road in the west,Yingpan Road in the north,Huangxing Middle Road in the east,and Zhongshan Road in the south,with a total area of about 20.86 hectares.In addition to the main Chao Zong street, it also includes streets with strong historical and cultural atmosphere, such as Beizheng Street, Lian Sheng street, Fuqing street, Yongqing Lane and Nanmu hall Lane. Among them,Chaozong Street and Beizheng Street,like Taiping Old Street,are key historical and cultural protection projects in Changsha. Chaozong Street starts from Xiangjiang Road in the east and ends at Huangxing North Road in the west.It is one of the only four ancient stone streets in Changsha.It is now about 400 meters long and 9 meters wide,making it the widest street in the old days.

Chaozong Street Historical and Cultural Block location is superior, surrounded by many famous scenic spots or historical buildings in Changsha, such as Wanda Plaza, Changsha children's palace, Wuyi Square, as well as Zhongshan Pavilion, Beizheng street Christ Church, Xichang street residential buildings and other historical buildings.

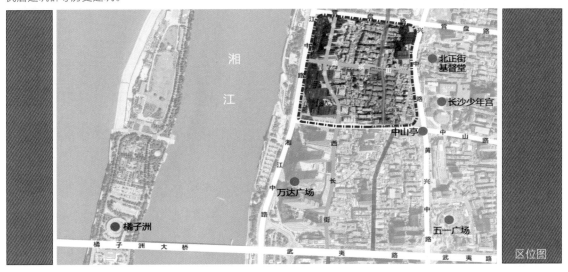

区位图

● 设计理念 Design philosophy

阅读历史，触摸时光

穿越不同年代，见证长沙城市演变过程。永清巷内各个时代建筑特色鲜明，同时有着历史悠久的行商文化记忆，故结合政府主导的整治开发模式，打造一条可以体现长沙不同时期建筑特色及长沙本土商业进化发展史的时空隧道。结合永清巷—寿星街的建筑年代分布规律，将永清巷部分分为 4 个组团：

Read history,feel time

Passing through different ages and witnessing the evolution of the city of Changsha. The architectural characteristics of each era in Yongqing Lane are distinctive and have a long history of business culture memory. Therefore, combined with the government-led renovation and development model, we will create a time tunnel that can reflect the architectural characteristics of Changsha and the development of local commerce in Changsha.Combining the chronological distribution law of buildings in Yongqing Lane-Shouxing Street, the Yongqing Lane is divided into 4 groups:

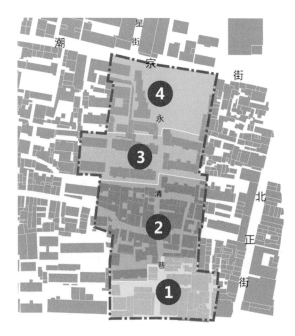

21 世纪后组团 ---- 现代集中复合商业
整改模式以拆除重建为主，建筑形式主要为新中式建筑，体现现代建筑风貌

新中国成立后 ---- 21 世纪前组团—现代个体商业
整改模式以保护、整治为主，建筑形式延续原有筒子楼，整治风貌的同时植入新功能

民国时期组团 ---- 民国行商
整改模式以保护为主，建筑肌理延续原有公馆建筑群，适当在这一区域植入开放空间

民国前组团 ---- 传统商铺
整改模式以拆除重建为主，建筑肌理取用连升街个体商户，体现民国前长沙老街的风貌特色

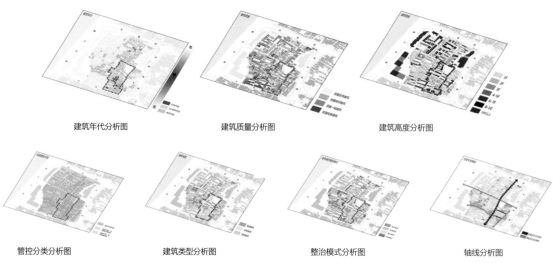

建筑年代分析图　　　　建筑质量分析图　　　　建筑高度分析图

管控分类分析图　　　建筑类型分析图　　　整治模式分析图　　　轴线分析图

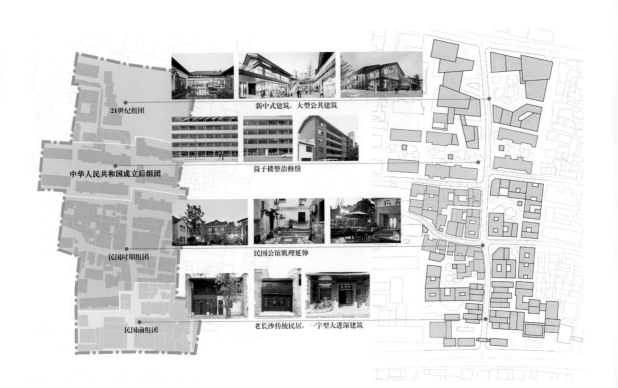

21世纪组团 — 新中式建筑，大型公共建筑

中华人民共和国成立后组团 — 筒子楼整治修缮

民国时期组团 — 民国公馆肌理延伸

民国前组团 — 老长沙传统民居，一字型大进深建筑

● **效果图 Effect picture**

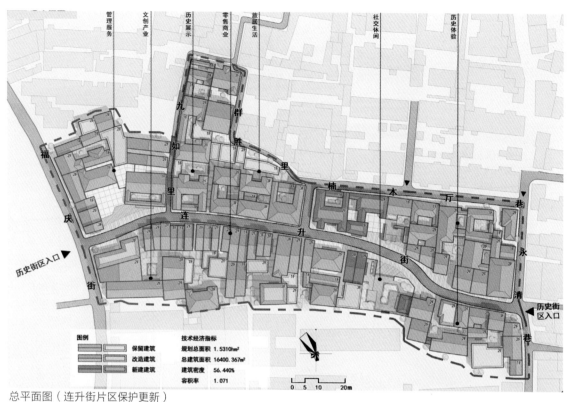

管理服务　文创产业　历史展示　零售商业　旅居生活　社交休闲　历史体验

九如里　群胜里　福庆街　连升街　楠木厅　木巷　永清巷

历史街区入口▶

历史街区入口▶

图例		技术经济指标	
	保留建筑	规划总面积	1.5310hm²
	改造建筑	总建筑面积	16400.367m²
	新建建筑	建筑密度	56.440%
		容积率	1.071

0　5　10　　　　20m

总平面图（连升街片区保护更新）

鸟瞰图（连升街片区保护更新）

设计题目： 长沙连升街历史文化街区保护与更新规划
指导老师： 向辉　孙亮
学　　生： 韩照伟

20

Design topic: Conservation and Renewal Planning of Historical and Cultural Block in Liansheng Street, Changsha

Instructors: Xiang Hui，Sun Liang

Student: Han Zhaowei

● 设计说明

连升街历史街区拥有丰富的历史文化和历史遗存，是长沙市历史文化街区的核心保护地段。在上位规划中，该场地属于政府主导开发的地块。本次设计以恢复连升街民国时期的历史文化风貌为主要目的，以空间设计和文化产业策划两个方面为切入点。规划通过文化转译、新产业置入、居民活力提升等设计手法丰富场地的文化内涵。通过整理街巷、流线串联、整合建筑肌理、置入公共空间等空间设计手法，吸引外来游客、促进原住民与外来游客的交流、提升地块的活力。

Design notes

Liansheng Street Historic District has rich Historical Culture and Historical Relics. It is the core protected area of Changsha Historic and Cultural District. In the Upper Planning, the site belongs to the plot of land developed by the government. The main purpose of this design is to restore the historical and cultural features of Liansheng Street during the period of the Republic of China, and the starting point is Space Design and Cultural Industry Planning. The plan enriches the Cultural Connotation of the site through design techniques such as Cultural Translation, Placement of New Industries and Improvement of Residents' Vitality. By sorting out streets and lanes, connecting streamline, integrating architectural texture, placing public space and other space design methods, we can attract foreign tourists, promote the communication between indigenous people and foreign tourists, and enhance the vitality of the plot.

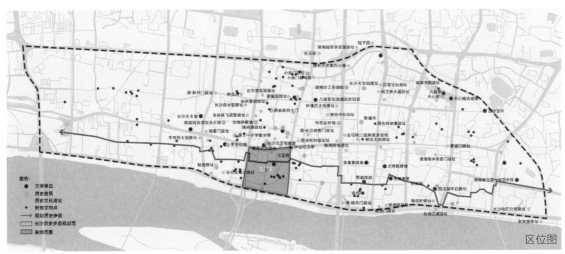

区位图

● 分析图　Diagram

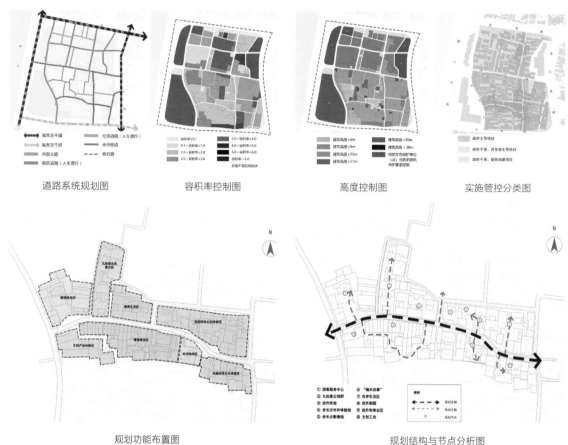

道路系统规划图　　　　　容积率控制图　　　　　高度控制图　　　　　实施管控分类图

规划功能布置图　　　　　　　　　　规划结构与节点分析图

● 立面效果图　Facade rendering

连升街道路北立面图

连升街道路南立面图

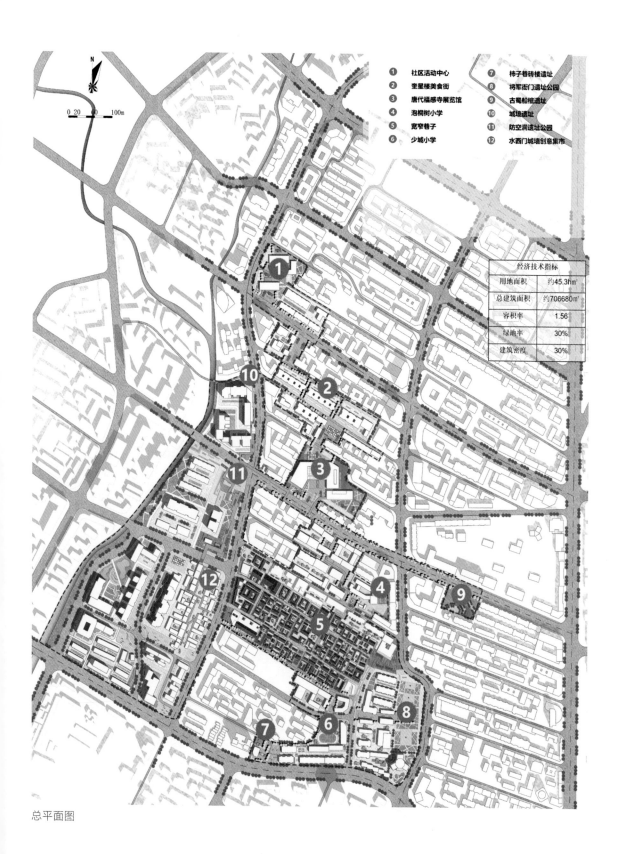

图例	
① 社区活动中心	⑦ 柿子巷砖楼遗址
② 奎星楼美食街	⑧ 将军衙门遗址公园
③ 唐代福感寺展览馆	⑨ 古蜀船棺遗址
④ 泡桐树小学	⑩ 城墙遗址
⑤ 宽窄巷子	⑪ 防空洞遗址公园
⑥ 少城小学	⑫ 水西门城墙创意集市

经济技术指标	
用地面积	约45.3hm²
总建筑面积	约706680m²
容积率	1.56
绿地率	30%
建筑密度	30%

总平面图

设计题目： 开街引流，古城复联——成都市少城片区城市更新设计
指导老师： 孙亮
学　　生： 彭雅欣

21

Design topic: Urban Renewal Design of Chengdu Shaocheng District

Instructor: Sun Liang

Student: Peng Yaxin

● 设计说明

少城片区位于四川省成都市青羊区天府广场西侧，处于城市核心区，是成都"三城相重"历史格局的重要组成部分。调研阶段以少城片区为研究对象，从宽窄巷子与少城其他区域的认知度不匹配为切入点，围绕社会、空间、经济三个层面开展调研，深入分析了其认知方面存在差异的原因，得出宽窄巷子与少城其他片区存在着弱关联性的问题。在设计阶段，从两者之间的弱关联性为切入点，以"连接"作为设计概念，通过连接历史、连接人群、连接业态三种途径，达到增强两者关联性的目的，最终使得少城能够协同发展，少城文化得以传播，人群之间能够互惠共享。

Design notes

Shaocheng District is located in the west of Tianfu Square, Qingyang District, Chengdu city, Sichuan Province. In the core of the city,it is an important part of the historical pattern of "three cities overlapping each other" in Chengdu. In the research stage, the small city area is taken as the research object, and the recognition mismatch between wide and narrow alleys and other areas of little City is taken as the entry point. The research is carried out from three aspects of society, space and economy, and the reasons for the difference in cognition are deeply analyzed. It is concluded that there is weak correlation between wide and narrow alleys and other areas of little city. In the design stage, starting from the weak correlation between the two, the design concept of "connection" is taken as the starting point, and the purpose of enhancing the correlation between the two is achieved through connecting history, connecting people and connecting business format. Finally, the collaborative development of the can be achieved, the culture of the Shaocheng District can be spread, and the people can share mutually.

成都少城空间格局变迁分析   图片来源：张霜霜"再见少城"—成都"少城"片区城市空间及其变迁研究 [D]. 成都：西南交通大学 ,2014.

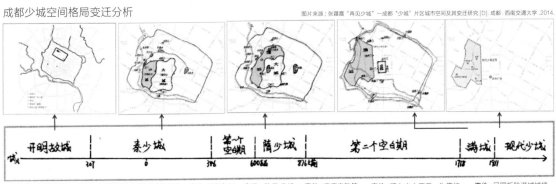

● 综合分析 Comprehensive analysis

人群的行为特征

(表2-1)　□ 居民　■ 游客

类型	具体行为	描述	现场照片	行为空间	行为主体	行为时间	驻留时长
生活性活动	跳广场舞	利用商场前的集散广场活动			居民	8:00-10:00	
	打羽毛球	利用街道空间进行体育活动			居民	10:30—11:30	
	聊天闲读	利用街道空间或改造街道空间进行闲聊			居民	8:30—11:00 13:00—17:00	
	棋牌活动	在棋牌室内或者利用室外公共活动空间			居民	13:00-17:30	
	接孩子	放学时间在校门口聚集大量接孩子的家长			居民	16:00-17:00	
	晒太阳	天气好的下午人们热衷于在街道旁休憩			居民&游客	14:00-17:00	
	散步跑步	在人民公园进行遛弯、健走、散步等锻炼			居民&游客	14:00-17:30	
商业性活动	采耳	在采耳馆或者街头的采耳摊进行采耳			居民&游客	全天	
	吃饭	藏匿在街头巷尾的苍蝇小馆是人们的最爱			居民&游客	11:00—14:00 17:00—20:00	
	品茶	茶馆随处可见，人们喜欢在其中消磨时间			居民&游客	10:00-17:30	
	打卡拍照	打卡网红景点店铺小吃，在特定场景拍照			居民&游客	全天	

空间自相关分析　2020.12.3人群分布热值　2020.12.3聚类与异常值

空间自相关分析　2021.3.4人群分布热值　2021.3.4聚类与异常值

在社会层面，通过实地调研对人的行为特征进行归类分析，并将少城地区划分成164个网格，对同时段各网格的人群聚集程度和不同时段几条重点街巷的人群聚集程度分别进行数据搜集，通过GIS的聚类分析可知，无论哪个时间段，宽窄巷子的人群聚集程度都远远高于其他街巷。并通过三大人群类型行为轨迹的跟踪模拟可以得出，不同类型的社群之间活动轨迹基本无交叉，游客、居民、商家之间的交往受到阻碍。

At the social level, through field research, the behavioral characteristics of people are classified and analyzed, and the Shaocheng District are divided into 164 grids. Data are collected on the degree of crowd aggregation in each grid at the same time and the degree of crowd aggregation in several key streets and alleys at different times. Through the cluster analysis of GIS, it can be known that no matter in which time period, the degree of crowd gathering in wide and narrow lanes is much higher than in other streets. And through the tracking simulation of the behavior trajectory of the three crowd types, it can be concluded that the activity trajectory of different types of communities basically does not cross, and the communication between tourists, residents and businesses is hindered.

历史遗存

□ 宽窄巷子　■ 历史遗存

地点	历史遗迹	年代	保护等级	所处位置	现存状态	现场照片
少城其他区域	西马棚民居	清代	优秀传统风貌建筑		改造为火锅店	
	四号工厂青年旅社	中华民国	不可移动文物		改造为青年旅社	
	福感寺遗址	隋唐	勘测中的遗址		荒地，现为临时停车场	
	民国励志社	中华民国	成都市近现代优秀建筑		现为政府办公处	
	古城墙	清代	市级文物保护单位		中间C门路段的古城墙改造为街角公园，其余两处均未找到遗迹	
	成都画院	清代	省级文物保护单位		正在修缮，现为公益性专业艺术机构	
	古蜀船棺遗址	先秦	全国重点文物保护单位		未找到遗址确切所在地，疑似为社区花园	

历史遗存

□ 宽窄巷子　■ 历史遗存

地点	历史遗迹	年代	保护等级	所处位置	现存状态	现场照片
少城其他区域	柿子巷19号民居大门	中华民国	不可移动文物		保留	
	柿子巷民居	中国民国	市级文物保护单位		正在修缮	
	长顺街118号砖楼	中华民国	区县级文物保护单位		保留，改造为茶楼	
	将军衙门	清代			已消失历史文化资源点	消失，改为金河宾馆，周边作为停车场和民居
	少城综合学校旧址	清代	区县级文物保护单位		现改造为餐馆	
宽窄巷子	宽窄巷子民居	清朝民国	不可移动文物		民居或改造成商铺	
	宽窄巷子民居	清朝民国	优秀传统风貌建筑		改造成商铺	

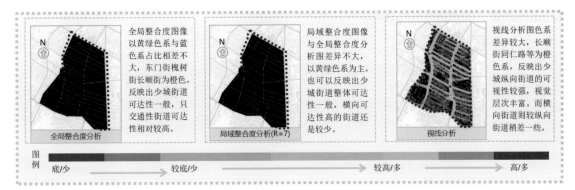

全局整合度图像以黄绿色系与蓝色系占比相差不大,东门街槐树街长顺街为橙色。反映出少城街道可达性一般,只交通性街道可达性相对较高。

全局整合度分析

局域整合度图像与全局整合度分析图差异不大,以黄绿色系为主。也可以反映出少城街道整体可达性一般,横向可达性高的街道还是较少。

局域整合度分析(R=7)

视线分析图色系差异较大,长顺街同仁路等为橙色系,反映出少城纵向街道的可视性较高,视觉层次丰富。而横向街道则较纵向街道稍差一些。

视线分析

图例 | 底/少 ——————→ 较底/少 ——————————→ 较高/多 ———→ 高/多

在空间层面,对少城的历史遗存和交通联系进行了分析。通过对历史遗存的梳理可以发现少城遗留下来的历史遗迹较少,大多都集中在宽窄巷子及其周边地区,且其中绝大多数都没有得到良好的利用,处于闲置的状态。而基于空间句法原理对少城的路网结构进行分析可以得知,少城街道整体的可达性一般,交通性质的几条主要干道可达性较强;纵向街道的视觉层次较丰富,横向街道引导性强,较为单调;南北地块的连接度不高。

At the spatial level, the paper analyzes the historical relics and traffic connections of the Shaocheng District. Through the sorting of historical remains, we can find that there are few historical relics left in Shaocheng District, most of which are concentrated in wide and narrow alleys and surrounding areas, and most of them are not well used and lying idle. Based on the principle of space syntax, it can be seen from the analysis of road network structure of little city that the overall accessibility of streets in little city is not good, and the accessibility of several main roads of traffic nature is strong. The visual level of longitudinal street is rich, while the horizontal street is strong and monotonous. The connection between the north and south plots is not high.

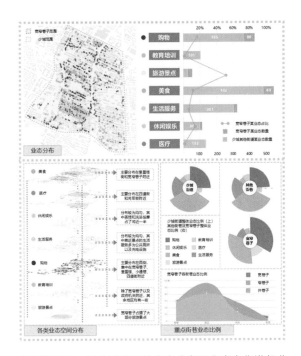

业态分布

各类业态空间分布

重点街巷业态比例

而通过少城片区的业态变化可以发现宽窄巷子最初的规划并没有考虑其他街巷,随着奎星楼街的居民率先自发破墙开店,成功将宽窄巷子的人流引入奎星楼街,才有了之后的小通巷、泡桐树街的更新改造,可见将宽窄巷子的人流引入其他街道的需求是存在的。

On the economic level, through the analysis of the distribution and change of the formats of business in Shaocheng District, it can be seen from the distribution of formats that both formats mainly focus on shopping and food, but the formats of wide and narrow lanes mostly serve tourists. Other streets and lanes in shaocheng mostly serve residents.
Through the change of the formats of Shaocheng width alley can be found in the initial planning failed to consider other streets,as KuiXing floor residents take the lead in opening stores by demolition, and successfully introduce the people from the wide and narrow alleys to KuiXing building street, Therefore leading the renewal of Xiaotong Alley and Paotongshu Street,so we can know that the need to introduce the people there to other streets does exist.

在经济层面,对少城片区的业态分布和业态变化进行分析,从业态分布可以得知两者的业态都以购物、美食为主,但宽窄巷子的业态多服务于游客,少城其他街巷的业态多服务于居民。

● 设计策略　Design strategy

设计方案从弱关联性作为切入点，以加强宽窄巷子与少城其他街巷的联系性为目标，将"连"作为概念，通过修补空间肌理、强化历史轴线、建筑更新改造等方式"连接"历史；通过完善步行体系、增强纵向联系、营造公共空间等方式"连接"人群；通过业态更新置换、新型功能植入、完善配套设施等方式"连接"业态，达到古城"复联"的目的。

Starting from weak correlation, the design aims to strengthen the connection between wide and narrow alleys and other streets and alleys in Shaocheng District. The concept of "connection" is taken as a concept to "connect" the history by repairing the spatial texture, strengthening the historical axis, and building renewal and transformation. "Connect" people by improving the walking system, enhancing vertical connections and creating public spaces; By means of updating and replacing formats, implanting new functions and improving supporting facilities, "connecting" formats, the purpose of "reconnecting" ancient cities is achieved.

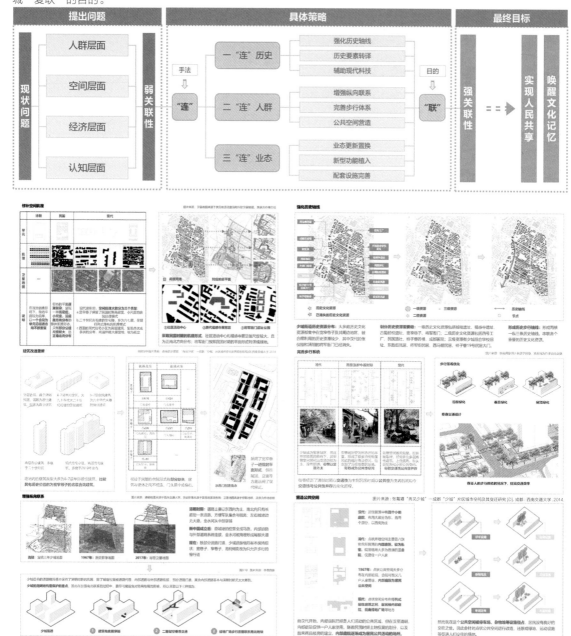

● 鸟瞰图　Aerial view

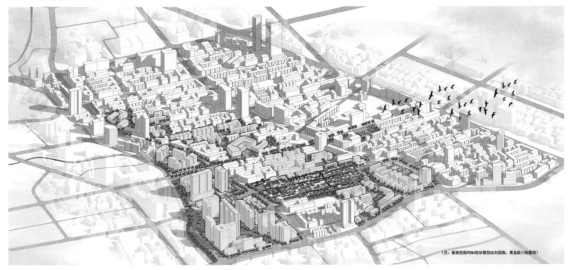

(注:基地范围内现状模型由刘国图、黄龙超小组提供)

● 节点设计　Node design

唐代福感寺展览馆　　　　　　　　　　城墙感应区　　　　　　防空洞遗址公园

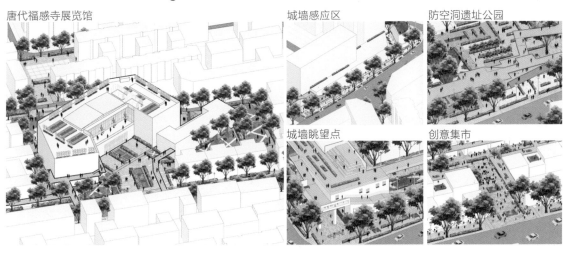

城墙眺望点　　　　　创意集市

城墙遗址公园

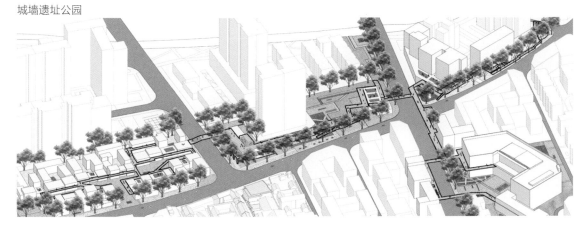